Live Better, Longer

The Science Behind the
Amazing Health Benefits of OPCs

RICHARD A. PASSWATER, PH.D.

Basic Health
PUBLICATIONS, INC.

The information contained in this book is based upon the research and personal and professional experiences of the author. It is not intended as a substitute for consulting with your physician or other healthcare provider. Any attempt to diagnose and treat an illness should be done under the direction of a healthcare professional.

The publisher does not advocate the use of any particular healthcare protocol but believes the information in this book should be available to the public. The publisher and author are not responsible for any adverse effects or consequences resulting from the use of the suggestions, preparations, or procedures discussed in this book. Should the reader have any questions concerning the appropriateness of any procedures or preparation mentioned, the author and the publisher strongly suggest consulting a professional healthcare advisor.

Basic Health Publications, Inc.
28812 Top of the World Drive
Laguna Beach, CA 92651
949-715-7327 • www.basichealthpub.com

Library of Congress Cataloging-in-Publication data is on file
with the Library of Congress.

ISBN-13: 978-1-59120-209-7

Editor: Cheryl Hirsch
Typesetting/Book design: Gary A. Rosenberg
Cover design: Mike Stromberg

Printed in the United States of America

10   9   8   7   6   5   4   3   2   1

Contents

Introduction

This book will help you be healthier and live better longer. Most people, including many physicians, have yet to understand that most of the non-germ diseases— the diseases that occur as they age— have a common cause, and it's not time. Instead, it's free radicals and it is exciting to realize there is actually one safe, simple, natural way to block that single cause and prevent the development of the more than sixty diseases that stem from it. I, and hundreds of other scientists, have found that free radicals can be blocked by a previously neglected family of natural nutrients contained in the dietary supplement OPCs.

Free radicals are involved in the diseases and disorders commonly associated with growing older, such as arthritis, cancer, or heart disease, and you can reduce your risk of these diseases related to aging with OPC supplementation. Yes, there is no doubt of this among research scientists in the fields of free-radical pathology and aging, but because the science is still too new for everyone in the clinical field to have been taught the new information, many health professionals still think of these diseases either as being the result of time, age, or other causes. It takes years and often decades for this information to reach the clinical field.

The fact that many diseases can be caused by one factor should not be that hard to comprehend in this post-Pasteur age. When Pasteur was trying to convince scientists and physicians that many diseases were caused by germs, he was ridiculed and scorned. How could the same thing that causes diarrhea or malaria cause a sore throat? Today, the fact that germs are the cause of many diseases is totally accepted, even though a different germ (bacterium or virus) causes each different disease. The body of knowledge about the causes of diseases has now been extended, and it is understood that over time this free-radical family of very reactive chemicals is involved in many debilitating and fatal diseases.

How can cancer, heart disease, and arthritis have this common cause, you wonder? After all, according to common wisdom, heart disease is caused by cholesterol, cancer is caused by genes, and arthritis is caused by wear and tear. This book will present you with overwhelming evidence that proves otherwise—evidence that is widely accepted by the many scientists who study this process and by the thousands of physicians who have used this information to improve the health, longevity, and quality of life for thousands of their patients. The mainstream, however, is just beginning to understand this process, and millions of people out there need to know this information.

Just by following the story of how these discoveries were made, you will be able to understand this common cause and learn how to prevent it or significantly slow it down. The story starts with some important but not dramatic discoveries and then builds rapidly with the most useful and exciting discoveries coming later as research continues. The beauty is that you can put these discoveries to practical use immediately.

There is even more good news. Not only can the newly understood nutrients protect against the initiation of these diseases, they can also help those who already have these diseases and disorders.

I want to introduce you to the health benefits of this family of versatile antioxidant nutrients of the oligomeric proanthocyanidins family of bioflavonoids called OPCs. Many years of research have shown they can prevent and treat many diseases. In the following chapters, I will discuss the field of free-radical pathology and elaborate on specific research with specific diseases.

Free-radical pathology is the study of how these very reactive chemical species called "free radicals" are involved in many disabling and fatal diseases. Free radicals and their close reactive chemical relatives are produced both in the body as part of the life process and additionally as byproducts of pollutants and environmental factors, such as natural radiation, including sunlight. I will discuss free radicals and the harm they cause in the first chapter, but for now, all you need to know is that these chemicals can harm the functioning of body components and cause disease and disorder. It is also useful to understand these harmful free radicals *can* be controlled and minimized by certain antioxidant nutrients.

At this time, the scientific community is particularly interested in the OPC family of very powerful antioxidant nutrients. These nutrients have always been present in foods in small quantities, but were not well studied until recently. Once their antioxidant power was realized, scientific interest in them grew rapidly.

OPCs have a number of key actions. They can:

- Boost immunity;

- Ease hay fever and allergy symptoms;

- Help keep skin smooth and youthfully flexible;

- Help protect against complications of diabetes, such as retinopathy;

- Help relax blood vessels, thereby improving blood circulation and helping to normalize blood pressure;

- Improve circulation and help keep blood cells slippery so they don't cause the blood clots that result in heart attacks;

- Improve learning ability and memory retention;

- Protect against the dangerous free-radical molecules, which speed up the aging process and set the stage for cancer, heart disease, and more than sixty other diseases;

- Reduce inflammation and help restore and maintain joint flexibility;

- Reduce the effects of stress;

- Reduce the risk of cataracts;

- Strengthen blood vessel walls, protect the linings of blood vessels, and reduce edema (swelling).

The discoveries by the scientists in this field are important to your health, wellness, and longevity. Yet, it has taken a long time for their information to be put to practical application to help people. I have been involved in free-radical pathology since 1960 and I have tried to speed up the process of transferring information from the scientific literature to the public by writing over forty books for the general public about how nutrients can benefit them.

It does little good for scientists to devote their professional lives to studying how certain nutrients can help people have a better quality of life if the research never gets put to practice use. The importance to you is to be able to use the information now, rather than having to wait decades for it to emerge from the literature. There are few patents on nutrients, so pharmaceutical companies that depend on patent protection of their products to recover their expenses on research and product education do not spend any funds to educate the medical profession and public about the benefits of nonpatent-protected nutrients.

To counter this, I am presenting my recent research, and that of my colleagues and friends, which contains new information for you to evaluate while it can do you the most good. I will also review some of my earlier research that is now widely known and accepted. I have found that if I publish important health information in popular book format, I can bring the benefits of my research to people much quicker than waiting for it to fight the uphill battle against current scientific beliefs. Although there are many physicians who have put these scientific discoveries into practice and can verify the results on a clinical basis—including the anti-aging experts such as Drs. Ronald Klatz and Robert Goldman, and many members of the American College for the Advancement in Medicine (ACAM) and the American Association of Anti-Aging Medicine (A4M)—with the information in this book, you can now learn for yourself how to put these discoveries to use to improve your health.

CHAPTER 1

The Battleground for Health:
OPCs versus Diseases Caused by Free Radicals

How can OPCs help protect against so many diseases and disorders, you ask? You should be skeptical of such broad health claims. Fortunately, there is a scientific explanation and clinical substantiation for the claims about OPCs. The information you are about to read is still too new for many health experts to have learned through their traditional educational channels. Practicing physicians normally do not do biochemical or nutritional research, and after they complete their primary medical education, they receive continuing education only in their specialty channels. The exceptions are those physicians and healthcare providers who specialize in complementary medicine.

The information presented in this book is the latest research on OPCs. Before getting started, I believe it is important for you to know something of my own background in antioxidant research, as it represents my life's professional work. I began conducting laboratory experiments with antioxidant nutrients in 1960, and was the first scientist to study the role of antioxidant combinations on health and lifespan. This research led to my discovery of how certain combinations of antioxidant nutrients were synergistic, meaning that when antioxidants work together, they have an effect even greater than the sum of their individual effects, and could slow the aging process and reduce the risk of various diseases. In 1969, I applied for patents on this process.

In 1970, in Toronto, I presented evidence to the Gerontological Society's Annual Scientific Congress that antioxidant nutrients offered a practical means of increasing human lifespan. I was the first to show that practical dosages and combinations of antioxidant nutrients can increase the lifespan of laboratory animals (*Chemical & Engineering News*, 1970). Additional details of my research were published in *American Laboratory* and *International Laboratory* in 1971.

At about this same time, reports of my research began to appear in publications for the general public. In 1970, *Ladies Home Journal* published an article about my research by Patrick McGrady, Jr. Another article appeared in 1971 in *Prevention* magazine. The first times the terms "antioxidant nutrient" and "free radical" ever appeared in the lay press were in these early articles.

Eventually, I received the first in a series of patents on antioxidant synergism (U.S. Patent No. 6,090,414). I wanted to put this information to practical use to help people right away so I wrote *Supernutrition: MegaVitamin Revolution* (Dial Press, 1975), and began lecturing at the American College for the Advancement in Medicine (ACAM) and other medical organizations interested in nutrition as well as medicine. Thousands of physicians have put my antioxidant discoveries to good practice with their patients. But the discoveries have continued, and there is now new information to report. It is important that you know about these early discoveries so that you can understand how the disease relationship between free radicals and antioxidants works. But it is the later discoveries about powerful and versatile antioxidants of the oligomeric proanthocyanidins class of bioflavonoids called OPCs that may be the most important, as they are very effective.

Early Discoveries about Free Radicals and Antioxidants

Let's look now at how this one dietary supplement can help protect against so many diseases and disorders. While the answer lies mostly in the fact that there is one common link among all diseases, there is more to it than that.

Part of the answer is that OPCs are not just one nutrient. They are a group of very powerful antioxidant nutrients within the same family, each with their own diverse actions. Most of the compounds act chiefly as antioxidants, while others block the actions of undesirable compounds in the body. This will be explained shortly.

Another part to the answer is that some nutrients, such as antioxidants, affect many body systems, and thus are factors in preventing many diseases. Antioxidants are involved in reducing the risk of more than sixty diseases.

As mentioned in the Introduction, OPCs have a number of key actions. The five most important actions of OPCs are their ability to:

1. Terminate free radicals and protect cells;

2. Enhance immune function and increase protection against many infections;

3. Bind to the skin proteins collagen and elastin to protect tissue and seal leaky capillaries;

4. Improve the functioning of blood vessels from the large arteries to the microcirculatory capillaries; and

5. Inhibit tiny blood cells from sticking together, thereby preventing circulation problems from heart attacks to strokes to deep vein thrombosis.

To fully understand the unique health benefits contained in OPCs, it's necessary to begin the story of these discoveries with a little background on free radicals. Free radicals are harmful molecules that damage the body. As will be explained in later chapters, free radicals are involved in cardiovascular diseases and other chronic degenerative diseases. They accelerate the aging process, and as a result, are indirectly involved in decreasing your defenses against germs. Arthritis, Alzheimer's disease, and Parkinson's disease are also linked to damage by free radicals. Studies indicate that antioxidants likely reduce the risk of these diseases.

You can think of free radicals as biological terrorists. Quite simply, they can be bad for your health. In chemistry, atoms that often are found grouped together are called a "radical." This group or radical generally stays together during a chemical reaction and can be transferred from molecule to molecule.

Sometimes during very high-energy chemical reactions, radicals can have an electron pulled away, causing the group to temporarily break free from the molecule. When this happens, it is called a "free radical." While this unstable, high-energy fragment is free, its energy forces can attract an electron from other molecules. A free radical can pull an electron from most biological compounds, thus restoring its original electron content, but causing the other compound that has lost an electron to itself become a free radical.

Some healthcare practitioners confuse free radicals with ions. An ion is simply a molecular fragment having a charge—either positive or negative. A free radical may or may not have a charge. The distinguishing feature of a free radical is that it contains one or more "unpaired" or "lone" electrons. It is not important for you to understand the difference, but scientists and health professionals should. They may wish to consult the more full description of free radicals and reactive species in the inset (opposite).

There are a number of known free radicals that occur in the body, the most common of which are oxygen-derived free radicals, such as superoxide radicals and hydroxyl radicals, hypoclorite radicals, hydrogen peroxide, and various lipid peroxides. They may be formed from exposure to environmental toxins like industrial pollution, household chemicals, and cigarette smoke, from the synthetic chemicals that are added to our water and food, and from everyday metabolic pathways that occur in the body to produce energy.

This free-radical reaction can perpetuate until a key biological molecule becomes permanently damaged. Scientists have estimated that each cell in your body (and you have billions of cells) suffers 10,000 free-radical "hits" each day. The amount of damage depends on how well the cell is protected by antioxidants. The higher your levels of antioxidants, the greater the amount of protection.

Free Radical Damage

Free radicals can damage the body's proteins and cell membranes,

A DESCRIPTION OF FREE RADICALS FOR SCIENTISTS AND THE SCIENTIFICALLY CURIOUS

Free radicals are harmful chemical species normally produced in the body as a byproduct of utilizing oxygen to produce energy. The body purposely creates certain free radicals for specific purposes, but most free radicals in the body are formed unintentionally and are undesirable agents that can do damage that leads to arthritis, cancer, heart disease, accelerated aging, and diseases and disorders associated with aging.

Simply, a free radical is an active part of a molecule. Another simple description that holds true for the vast majority of biochemical situations is that a free radical is merely a chemical species with an odd number of electrons. All chemical species having an odd number of electrons are free radicals, but there are chemical species that have an even number of electrons that are also free radicals, although they are rarely encountered in biochemistry. A slightly better description is that a free radical is a molecule or atom containing one or more lone or unpaired electrons. As explained in the text, this is not to be confused with an ion, which is a chemical species with an imbalance between the numbers of negatively charged electrons and positively charged protons. Free radicals can be electronically neutral or charged. If a free radical is charged, it is called a "radical ion." The energy of interest in free radicals is magnetic rather than electrical.

Atoms that are grouped together more or less as a unit in a molecule can be called a radical. This group or "radical" generally stays together during a chemical reaction and can be transferred from molecule to molecule. However, when speaking of "free radicals," radical simply refers to the remainder of an atom or molecule after an electron has been removed from a pair of electrons in an electronic orbital.

Electrons travel about the nucleus of an atom or molecule in paths similar to the manner in which planets travel around the sun. Atoms and molecules have various layers or shells of electrons orbiting a nucleus. Electronic shells and orbitals are merely regions in which there is a high probability of finding an electron. These shells and orbitals are determined by the structure of the atom or molecules. Filled atomic electronic shells hold two, eight, eighteen, and thirty-two electrons, and electronic orbitals hold a maximum of two electrons.

Traveling within these atomic or molecular electronic shells are one or more electronic orbitals that can hold a pair of electrons. Electrons also spin about on their axis like the earth spins on its axis to give us day and night. Electrons found in atoms and molecules tend to exist in pairs. Normally, two electrons are coupled or paired in an orbital with their spins in opposing directions. By convention, these spins are described in quantum mechanics as being either "up" or "down." Since the spins of electrons normally (at standard conditions in what is termed the "ground state," which is usually a "singlet" state) are in opposing directions, their magnetic vectors cancel each other. The opposing spins of these paired electrons result in a lower energy state than resulting from other groupings. When an electron is by itself in an orbital, there is extra energy available due to its spin.

When there is a lone electron, the radical will have a magnetic vector and is said to be "paramagnetic."

The reason that the simple definition given earlier is accurate is that all atoms and molecules that have an odd number of electrons must have at least one electron that is in an orbital by itself. However, this simpler definition doesn't take into account such things as *di*-radicals, which are chemical species having two lone or unpaired electrons, and atoms of many elements, which are free radicals because they contain lone electrons in orbitals even though the total number of electrons may be even. Transition metal complexes that have one or more unpaired electrons associated with the metal are not generally referred to as free radicals.

All chemical species with an odd number of electrons are free radicals, but there are some free radicals that have an even number of electrons, but generally speaking, those exceptions are not the free radicals generally dealt with in biochemistry.

The simplest free radical is the hydrogen atom as it has only one electron. The hydrogen molecule has two electrons and is not a free radical. Other common free radicals include the oxygen diatomic molecule (O_2), superoxide anion radical ($O_2\bullet^-$), hydroxyl radical (HO•), alkoxyl radical (RO•), and the peroxyl radical (ROO•).

FREE RADICAL PROPAGATION

Being that free radicals have higher energy due to their unpaired electron, they have a tendency to be reactive with other molecules. This reaction is simply described as "an effort to regain its lost electron and become stable." Because of this, most free radicals are unstable and very reactive. Thus, free radicals can be harmful to body components. Most free radicals call pull an electron from most biochemical compounds. It has been estimated that each cell in the body is hit by approximately 10,000 free radicals every day.

A few free radicals are stable molecules. Nitric oxide, a chemical compound naturally found in the body, is an example of a free radical that is a stable molecule. Nitric oxide appears to be extremely important in maintaining normal blood pressure, cholesterol, and heart function.

When a free radical reacts with another compound, it generally removes an electron from that compound, which, of course, leaves it with an odd number of electrons and has now become a free radical. This series of reactions can continue until the reaction is terminated by a species that can alter electron spin or the last free radical formed is of low energy and has insufficient energy to continue the propagation. Antioxidants are compounds that quench free-radical reactions because their resulting free radicals have relatively low energy. The amount of damage occurred is dependent upon both the number of free-radical hits and the amount of protection by antioxidants.

OXYGEN FREE RADICALS

Most free radicals produced in the body are a byproduct of oxygen metabolism, that is, they are created as oxygen is utilized to produce energy from food components. Some oxygen free radicals are extremely harmful to the body. Actually, the oxygen molecule itself, a diatomic molecule consisting of two oxygen atoms, is a bi-radical. The ground-state diatomic oxygen molecule has two unpaired electrons, each located in a different non-bonding orbital. Thus, it has two odd electrons. But the oxygen molecule—which is actually two free radicals united—is not particularly dangerous because each of the spins of the unpaired electrons happen to be in the same direction. (Thus, the ground state of oxygen is the "triplet" state, rather than the singlet state.) If oxygen's unpaired electrons were spinning in opposite directions, oxygen would be a very reactive radical.

The first excited state of oxygen is the singlet state. Singlet oxygen, which is excited by 23 kcal/mol (kilocalories per mole), has all paired electrons. Thus, singlet oxygen is not a free radical, but is a reactive oxygen species (ROS).

In the metabolic process called the respiratory chain, in the mitochondria present in the cytoplasm of most living cells in which oxygen is used to create energy from the carbon and hydrogen in food while also producing carbon dioxide and water, some of oxygen reactions go astray. The carbon and hydrogen atoms in food lose electrons, thus increasing their valance state (oxidation number) and are thus "oxidized." Normally, oxygen is reduced by a two-electron step. This is called a "reduction" as it reduces the valance state (oxidation number) of oxygen.

Approximately 1 to 2 percent of the time, single-electron reactions occur creating a side reaction that produces very reactive oxygen radicals. There are several intermediate steps in this minor pathway that can reduce oxygen to water, but along the way they form some potentially dangerous free radicals and other reactive oxygen species.

When an electron is added to a diatomic oxygen molecule, it creates a superoxide anion radical, which has a negative charge. Once a superoxide anion radical is formed, free radicals may begin propagating throughout the body until they are terminated –that is, inactivated—either via the enzyme superoxide dismutase produced in the body or by any antioxidant.

weaken the cell's natural defenses, disrupt cells' DNA, and other essential body functions, that can lead to widespread biomolecular changes, leaving the body susceptible to many diseases. Free radicals can be the sole cause of a few diseases such as cataracts or some cancers, but more often are involved in the disease process by impairing the immune system and predisposing the body to diseases directly caused by other factors. Free radicals also may worsen the conditions and be antagonistic to the healing process.

First, let's look at the health conditions or diseases associated with free-radical damage that involve a particular organ or area in the body. They include:

- Blood cell disorders, including systemic lupus erythematosus and sickle-cell anemia;

- Brain disorders, including neurotoxicity, senile dementia, Parkinson's disease, Alzheimer's disease, stroke, and cerebral trauma from stroke;

- Gum diseases, including periodontitis;

- Eye disorders, including cataract, macular degeneration, ocular hemorrhage, degenerative retinal damage, diabetic retinopathy, and photic retinopathy;

- Gastrointestinal tract and liver disorders, including hepatitis, endotoxin liver injury;

- Heart and cardiovascular system disorders, including athero-sclerosis, heart attack, endothelial injury, vasospasms, and kidney disorders resulting from diabetic microangiopathy;

- Joint abnormalities, including rheumatoid arthritis;

- Lung disorders, including asthma, cigarette smoke-induced injury, emphysema, hyperoxia, bronchopulmonary dysphasia, cystic fibrosis, oxidant pollutant-induced injury, and acute respiratory distress syndrome (ARDS);

- Skin disorders, including melasma (chloasma), sunburn (solar radiation), burn (thermal injury), and psoriasis.

Then there are the conditions or diseases associated with free-radical damage that involve more than one organ or area of the body. They include:

- Aging, including disorders of "premature aging" and immune deficiency of aging;

- Cancer;

- Inflammatory-immune injury, including nutritional deficiencies, alcohol damage, and radiation injury;

- Reproductive disorders and ineffective sperm.

Antioxidants

Our best defense against free radicals and the development of the health conditions and diseases listed above are compounds called antioxidants. Oxygen is a very reactive element, which is why it rusts iron, promotes combustion, and supports the life process. Iron and iron-containing objects that are left out in air (which contains oxygen) rust, or as chemists say, "oxidize." The process by which things react with oxygen is called "oxidation." A substance that prevents or slows the oxidation process is called an "antioxidant." Antioxidants also protect other substances, such as living tissue, against damage caused by oxygen.

In the body, it's important for oxygen to be channeled into the right places and kept away from other places. For example, you don't want oxygen to react with vital cell components. This would damage them much like rust damages iron. In the body, unwanted oxidation of cell components can set the stage for aging, heart disease, cancer, and many other chronic degenerative diseases. Antioxidants sacrifice themselves to protect vital components.

To be considered an antioxidant, a compound must be so effective that a few molecules protect many, many other molecules by neutralizing bad molecules or fragments of molecules. Our bodies make some antioxidants. However, we are dependent on our diet to supply many antioxidants. Important antioxidant nutrients include vitamins, minerals, amino acids, and coenzymes.

Vitamins C and E are the two most "famous" antioxidants. Vitamin C prevents oxidation of water-soluble molecules and is a particularly effective antioxidant in the liquid-based areas of the body, including blood plasma, lung fluid, eye fluids, and in between cells. Vitamin E, on the other hand, is a fat-soluble vitamin most significantly present in the lipids of cell membranes. These vitamins reinforce and extend each other's antioxidant activity. Because the body cannot manufacture vitamin C or vitamin E, they must be obtained through the diet or in supplement form.

Minerals are not direct antioxidants, but several minerals can become vital components of antioxidant enzymes made by the body. Such minerals include selenium, which is needed to make the antioxidant enzyme glutathione peroxidase; iron, which is needed for catalase; and manganese, copper, and zinc, which are needed to make superoxide dismutase (incidentally, OPCs such as Pycnogenol® contain these minerals).

Sulfur compounds, such as the sulfur-containing amino acids cysteine and methionine, help the body produce the most common antioxidant within cells, glutathione.

Antioxidant coenzymes, such as nicotinamide adenine dinucleotide (NADH), coenzyme Q_{10}, and alpha-lipoic acid, can be made in the body as well as obtained in the diet.

Bioflavonoids are nutrients that can replace some of the need for the antioxidant vitamin C in some biochemical reactions and thus, save some vitamin C. Early research by Nobel Laureate Dr. Albert Szent-Györgyi suggested that bioflavonoids had additional properties that justified them being classified as vitamins. American scientists have rejected this view, but many European scientists believe that bioflavonoids are at least semi-essential and that the vitamin hypothesis merits further study.

Bioflavonoids

Bioflavonoids, often simply called flavonoids, are a class of thousands of beneficial compounds found in plants. The structures of these antioxidant compounds enable them to easily donate electrons to other molecules. This ability to donate electrons is characteristic of all antioxidants. There are thousands of bioflavonoids existing in nature. Scientists have identified over 4,000 of them, but they are sure that there are several thousand more yet to be identified.

Flavonoids are found in fruits, vegetables, nuts, seeds, grains, cacao, and in beverages such as tea and wine. Many flavonoids are pigments that provide several fruits with their blue and purple colors and some of the reds and emerald green.

In addition to their antioxidant properties, bioflavonoids have a host of other beneficial effects in the body. Studies have shown that bioflavonoids possess antiviral, anti-inflammatory, antihistamine, and even anticarcinogenic properties.

The importance of flavonoids is only just being realized by many nutritionists. In December 2006, Dr. Balz Frei, director of the Linus Pauling Institute and professor of biochemistry and biophysics at Oregon State University in Corvallis, published an article in *Free Radical Biology and Medicine* calling attention to the many biochemical pathways of flavonoids. Dr. Frei pointed out that flavonoids from fruits and vegetables are not only antioxidants when eaten but also contribute significant benefits by other mechanisms as well. "Flavonoids induce so-called Phase II enzymes [one of two families of enzymes] that also help eliminate mutagens and carcinogens, and therefore may be of value in cancer prevention," he said. "Flavonoids could also induce mechanisms that help kill cancer cells and inhibit tumor invasion."

Flavonoids have been attracting attention with a mounting body of science, including epidemiological (population) and laboratory-based studies that find the cancer-fighting potential of a number of different flavonoids, such as isoflavones, proanthocyanidins, and flavonols.

According to Dr. Frei, flavonoids act directly at low cellular levels or indirectly by increasing the levels of other antioxidants. Flavonoids may accumulate in tissues where they exert local antioxidant effects at very low concentrations by modulating cell signalling, gene regulation, angiogenesis, and other biological processes by non-antioxidant mechanisms, or they may act indirectly by increasing the levels of other antioxidants.

Additionally, the scientists involved with these studies noted that there is evidence that flavonoids increase the activation of existing nitric oxide synthase. This enzyme is responsible for the synthesis of nitric oxide, a chemical compound naturally found in the body that has positive effects on blood vessels, preventing inflammation, and lowering blood pressure.

Proanthocyanidins

Proanthocyanidins are naturally occurring plant metabolites widely available in fruits, vegetables, nuts, seeds, flowers, and bark. Good sources of proanthocyanidins include red wine, grape seeds, and pine bark. Sometimes proanthocyanidins are simply called "procyanidins." These bioflavonoids give color to plants, primarily the blue-violet and red pigment. Proanthocyanidins belong to the category of flavonoids known as condensed tannins, one of the two main categories of plant tannins.

It is not important for us to know this, but for the benefit of chemists who may be reading this, chemically speaking, proanthocyanidins are defined as high-molecular-weight polymers comprised of the monomeric unit flavan-3-ol—(+) catechin and (–) epicatechin. Oxidative condensation occurs between carbon C-4 of the heterocycle and carbons C-6 or C-8 of the attached A and B rings. The procyanidins B1–B4, characterized by the C4–C8 linkage, are the most common dimers, occasionally accompanied by corresponding C4–C6 linked isomers (B5–B8).

Even non-chemists may apreciate the fact that several scientists are now calling for epicatechins and their more complex family members such as OPCs to be classified as vitamins. Dr. Norman Hollenberg from Harvard Medical School and Brigham and Women's Hospital in Boston noted that people who consume large amounts of epicatechins have a 10 percent lower risk of cancer, diabetes, heart failure, and stroke. Dr. Hollenberg told *Chemistry & Industry* magazine that the importance of epicatechins in the diet is so great that it should be considered a vitamin.

Dr. Hollenberg's data was published in the March 2007 issue of the *International Journal of Medical Sciences*. "If these observations predict the future, then we can say without blushing that they are among the most important observations in the history of medicine," said Hollenberg. "We all agree that penicillin and anaesthesia are enormously important. But epicatechin could potentially get rid of 4 of the 5 most common diseases in the western world, how important does that make epicatechin? I would say very important."

Daniel Fabricant, vice president of scientific affairs at the Natural Products Association, a leading voice of the natural products industry, said: "The link between high-epicatechin consumption and a decreased risk of killer diseases is so striking, it should be investigated further. It may be that these diseases are the result of epicatechin deficiency."

Supporting studies find that flavonoid-rich diets are consistently associated with reduced heart disease incidence. As an example, a study of 34,489 postmenopausal women, published in the March 2007 issue of the *American Journal of Clinical Nutrition*, reported that a high dietary intake of several classes of flavonoids reduced the risk of mortality from cardiovascular disease (CVD), coronary heart disease (CHD) and stroke by between 10 and 22 percent. "This prospective study of postmenopausal women, with sixteen years of follow-up, is, to our knowledge, the first study that has been reported on total flavonoids and on seven subclasses of flavonoids," wrote lead author Pamela Mink from the Washington-based consulting firm, Exponent, Inc.

"Results from this study suggest that the intake of certain subclasses of flavonoids may be associated with lower CHD and total CVD mortality in postmenopausal women," said the researchers. "Furthermore, consumption of some foods that are high in flavonoid content or that are among the main sources of flavonoids in the diet of these study participants may have similar associations."

Oligomeric Proanthocyanidins (OPCs)

Now we finally get to the discussion of the class of nutrients that is so important to our health—the oligomeric proanthocyanidins, or simply OPCs. To understand where OPCs derive their name, consider a monomer as one bioflavonoid unit. Compounds consisting of two and three units (monomers) are called "dimers" and "trimers," respectively. Still larger compounds are present that consist of four to twelve units. These compounds are called "oligomers." "Oligo" simply tells us that the molecule consists of several smaller chemical units, but not so many as to form a large polymer. In this case, several units of proanthocyanidins are joined together to form a series of oligomeric proanthocyanidins, or OPCs.

OPCs are found in some foods in small amounts, however, the two main sources are pine bark extract (Pycnogenol®) produced from the bark of the French maritime pine tree, and grape seed extract, produced from the seeds, or pips, of the wine grape (*Vitis vinifera*). Red wine extract is also a powerful source of OPCs.

The Internet encyclopedia Wikipedia (www.wikipedia.com) describes OPCs as a class of flavonoids that acts as antioxidants in the human body. The web entry states further:

OPCs help protect against the effects of external and internal stresses like tobacco smoke, alcohol, and environmental toxins, such as radiation, pesticides, solvents, and heavy metals, as well as the stresses caused by normal body metabolic processes. These harmful substances trigger the formation of free radicals. OPCs eliminate free radicals and, as a result, help decrease the effects of these stresses. They can:

• Decrease risk of cardiovascular disease—OPCs decrease blood fat, emolliate (soften) blood vessels, lower blood pressure, prevent blood vessel scleroses, drop blood viscosity, and prevent thrombus formation (a type of blood platelet clot). Additionally, studies have shown that OPCs may prevent cardiovascular disease by counteracting the negative effects of high cholesterol on the heart and blood vessels.

• Strengthen and repair connective tissue—OPCs inhibit the body's enzymes that break down collagen. Cells are connected to one another with strong fibers called collagen. OPCs help collagen repair and rebuild these connections correctly, which can reverse damage done over the years by injury and free-radical attack. OPCs are taken as an oral cosmetic to help in the prevention of wrinkles.

• Act as a natural, internal sunscreen—Ultraviolet rays from the sun destroy up to 50 percent of your skin cells. OPCs reduce this amount to approximately 15 percent. Inhibiting the daily effects the sun's rays have on the skin is the best defense against the aging of our skin.

• Improve circulation—OPCs strengthen capillary walls, and thereby improve circulation. Capillaries are the smallest blood vessels in the body. They are responsible for supplying the tissues with a constant supply of blood for their nutrition and for repair purposes in the event of injury. They are also responsible for waste removal from the tissues. To do these functions properly, the capillaries must be healthy. (This is especially important for people with compromised circulatory systems, such as stroke victims, diabetics, arthritics, smokers, oral contraceptive users, and people with general cardiovascular insufficiencies.)

• Improve cognitive function—OPCs are one of the few antioxidant nutrients that readily cross the blood-brain barrier. This barrier normally protects the brain from free radicals circulating in the blood. This enables OPCs to fight free radicals in the vessels of the brain that in turn will help them remain healthy. This can result in increased mental acuity, a decreased potential for stroke, and the possibly of fighting senility.

• Moderate allergic and inflammatory responses—OPCs reduce histamine production. The release of inflammatory chemicals in the body (called histamines) initiates inflammation. OPCs inhibit the enzyme necessary for histamine production and prevent its formation, which further reduces inflammation. By reducing histamine production naturally, OPCs eliminate the need for manmade antihistamines and spare you of their side effects; therefore, proanthocyanidins are used in the treatment of allergies.

Synergism among OPCs

OPCs work synergistically, that is, they work together well. For example, when the three most powerful sources of OPCs—extracts of French maritime pine bark, grape seed, and red wine—are combined with two other important OPCs from bilberry and citrus— they have an effect even greater than the sum of their individual effects.

Rather than discuss each source of OPCs separately, as in a chapter on pine bark extract, a chapter on grape seed extract, and so on, I will discuss separate diseases and conditions and then explain what clinical studies have shown regarding their treatment with OPCs. But, first, here's a brief explanation of the principal sources of OPCs.

Pine bark extract (Pycnogenol) was the first source of OPCs discovered. This proprietary extract is a mixture of more than forty water-soluble and highly bioavailable nutrients extracted from the bark of a particular specie of maritime pine (*Pinus pinaster* Ait.) tree found in southwestern France. This extract contains a wide array of OPCs that serve many biochemical functions, especially powerful antioxidant activity. Many of the nutrients in Pycnogenol are bioflavonoids. Pycnogenol contains several classes of bioflavonoids, including monomers and oligomers. In addition, there are organic acids of the family often commonly called fruit acids. I have studied Pycnogenol since 1991 and have written four books and several articles on the research and clinical studies with it. Most of the research presented in this book is based on clinical studies with Pycnogenol. This OPC has been intensively investigated for more than thirty years, and as such there are many more clinical findings using Pycnogenol than any of the other OPCs.

Grape seed extract (*Vitis vinifera*) contains the OPCs gallic acid, epicatechin (EC), gallocatechin, epigallocatechin (EGC), epigallocatechin gallate (EGCG), and many other OPCs specific to grapes. Grape seed extract is also widely researched and found to be very beneficial. Grape seed extract and Pycnogenol have some similar health benefits and each has specific health benefits. Combining these two powerful OPC sources provides the benefits of both. I have often been asked, "Which is better, Pycnogenol or grape seed extract?" My answer is that both are important. Which is best apples or oranges? They each have important nutrients and you are better off eating both.

An excellent review of the health benefits of grape seed extract was published by Dr. Dabasis Bagchi and colleagues at Creighton University School of Pharmacy (*Toxicology*, 2000).

Red wine extract contains the OPCs that give red wine its characteristic color. These OPCs are found in high levels in the skin of red grapes but not in the seeds. Some researchers credit red wine to explain the "French Paradox"—the relatively low incidence of heart disease in France despite a fat-rich diet and a tendency to smoke. A glass of red wine contains averages about 45 milligrams (mg) of OPCs (Bagchi, 1999). In addition to other grape compounds, red wine extract contains the antioxidant resveratrol, which was called the "breakthrough nutrient" of the year in 2006 by several scientists. Resveratrol will be discussed later in the chapters on aging, heart disease, and cancer.

The herb **bilberry** (*Vaccinium myrtillus*), a member of the blueberry family, is famous for its use to sharpen the night vision of pilots in World War II. The OPCs are responsible for the deep blue color of bilberries, which is not only present in the skin of the berries, but also in the pulp of the berries. The OPCs in bilberry include cyanidin, delphinidin, malvidin, pelargonidin, and peonidin. Bilberries have long been used for vision and circulatory problems and have been prescribed since 1945 for diabetic retinopathy, a condition involving damage to the retina caused by complications of diabetes (Dupier, *Gale Encyclopedia of Alternative Medicine*).

Citrus bioflavonoids are a group of plant pigments found in citrus fruits. Dr. Albert Szent-Györgyi, a famed Hungarian researcher, found that citrus peel flavonoids were effective in preventing capillary bleeding and was the first to report the biological activity of flavonoids on capillary fragility (Szent-Györgyi, 1938). Important bioflavonoids in citrus include hesperidin, eriodictyol, rutin, quercetin, naringin, and many OPCs.

The benefit of selectively combining OPCs is antioxidant synergism. It is an example of the principle of antioxidant synergism I discovered in 1961 and that became the central theme of my patents. Let's take a look at how OPCs and other important antioxidants work together.

OPCs Are Synergistic with Other Antioxidants

In addition to OPCs' antioxidant compounds that have a direct protective factor, OPCs have an indirect protective effect by improving other antioxidants in the cells. OPCs can increase the levels of antioxidant enzymes produced by the body, as well as those antioxidants such as vitamins C and E that are derived from the diet. Dr. Wei and colleagues have shown that Pycnogenol can double the concentration of superoxide dismutase, catalase, and glutathione inside the cells (Wei et al., 1997).

One reason antioxidants work together synergistically is that some antioxidants can regenerate other antioxidants. For example, Pycnogenol can regenerate "used" or "spent" vitamin C, which in turn, can regenerate used vitamin E (Cossins et al., 1998). This means that Pycnogenol enables the sparse amounts of vitamin C and vitamin E found in most people's diets to function as if there were higher levels in diet. This is a result of recharging the spent vitamins instead of them being removed from the body.

When a vitamin E or vitamin C molecule comes into contact with a free radical, it donates an electron to the free radical and makes it a normal nonreactive molecule. This causes the vitamin E or vitamin C molecule to become a weak free radical. This weak free radical does no harm to the body, but since it has given up an electron itself, it can no longer stop other free-radical reactions by donating an electron. Thus, the vitamin E radicals or vitamin C radicals become useless. Usually the body simply destroys the inactive radicals by breaking them apart into smaller compounds to permit their removal from the body by way of the kidneys.

On the other hand, if the inactive vitamin C radical comes into contact with one of the bioflavonoids in Pycnogenol, it can be regenerated back into an active vitamin C molecule. Active vitamin C can also regenerate an inactive vitamin E radical back into an active vitamin E molecule. This effect of Pycnogenol is possible because the larger procyanidin molecules in Pycnogenol stabilize the lifetime of the inactive vitamin C radical so that it doesn't decompose and leave the body, but can last long enough to capture its missing electron from one of the many molecules of the procyanidins.

A Unique Antioxidant with Significant Health Benefits

OPCs are much more than powerful, multipurpose antioxidants. They also have strong anti-inflammatory, immune boosting, spasmolytic (anti-spasmodic), and anticoagulant (anti-blood clotting) actions. All of these actions when combined can give an OPC formula unique and significant health benefits.

As far as their antioxidant capabilities go, OPCs are very powerful antioxidants that are effective against many types of harmful free radicals. While some antioxidants such as vitamins C and E destroy several types of free radicals and other reactive oxygen species (ROS), a blend of OPCs having many different antioxidants destroys more types of free radicals and reaches more compartments of the body than any other antioxidant nutrient.

In addition, OPCs help balance and control the production of nitric oxide, which is a good compound in the artery linings and certain other areas, but can be a very harmful compound otherwise. I will discuss nitric oxide in the next chapter on heart disease.

Pycnogenol is so effective because of its powerful combination of OPCs. According to studies by Dr. Lester Packer and his colleagues at the University of California, Berkeley, in 1997, Pycnogenol is the most powerful antioxidant complex to be widely studied under identical laboratory conditions and reported in the scientific literature (Noda et al., 1997).

Results from the studies of Pycnogenol's antioxidant capabilities indicate that Pycnogenol may be the most powerful scavenger of oxygen free radicals and nitrogen free radicals, as well as other reactive oxygen species (ROS) and reactive nitrogen species (RNS) (Virgili et al., 1998). Pycnogenol has been also shown to be the strongest scavenger of hydroxyl free radicals and superoxide anion radicals among compounds and extracts tested (Noda et al., 1997).

When Pycnogenol was patented for its free-radical scavenging effect in 1987, it was described as being many times more powerful than vitamin E and vitamin C. This is true under certain conditions. The specific laboratory test used to make that comparison is called the nitrobluetetrazolium, or NBT, test and is only one measure of a compound's antioxidant and free-radical scavenging capabilities. This test, which is done in a water-based system, is certainly not a fair one solely to use to compare Pycnogenol's antioxidant properties with vitamin E's, as vitamin E is not soluble in water as are Pycnogenol and vitamin C. Conversely, it would not be fair to test Pycnogenol directly in a lipid (fatty) system.

When comparing antioxidants, several factors must be looked at. For example, one antioxidant may be better retained in one body organ or system than another. One antioxidant may be more efficient against one type of free radical then another. Many variables must be considered. The only fair way to compare antioxidants is to compare their profiles of actions against various free radicals.

Pycnogenol has been rated the best of the many antioxidants that Dr. Packer compared for their effectiveness against several free radicals that are present in the body. In those systems in which Pycnogenol has an effect—which includes many systems important to health such as blood vesssels, the eyes, blood platelets—he found Pycnogenol to be the most effective of all the nutrients tested. Part of Pycnogenol's uniqueness is that it can work in liquid-based areas of the body like vitamin C and in fat-based areas, where it restores vitamin E and other fat-soluble antioxidants that increase antioxidant activity in fatty systems.

In a more biologically relevant setting to humans, Dr. M. Chida and his colleagues carried out a study to compare the ability of different antioxidants to protect retinal lipids (fats) from damage by free radicals. The researchers found that Pycnogenol was far more potent than vitamins C and E, alpha-lipoic acid, coenzyme Q_{10}, and grape seed extract (Chida et al., 1999).

The point is, there may not be just one antioxidant nutrient that works best in all the systems—but Pycnogenol appears to be the most powerful in the most systems, especially in the systems of major importance, and it definitely should be in everyone's defense arsenal against free radicals.

One important reason why a blend of OPCs appears to be more effective than a single antioxidant compound is that the mixture contains various size molecules, which help these antioxidants reach various parts of the body.

Pycnogenol's complex of antioxidants provides compounds of varying sizes that can function effectively in different regions of the body over varying periods of time. The larger procyanidins in the mixture are very effective in the bloodstream, whereas the smaller flavonoid molecules and organic acids can readily enter cell interiors. The large oligomeric procyanidins have several points in their molecules that can annihilate free radicals. Vitamin E in contrast has only one such point. These free-radical annihilating points are called phenolic groups, and vitamin E is a monophenol whereas OPCs are polyphenols.

In addition, the various types of antioxidant compounds in OPCs make them multipurpose intracellular and extracellular antioxidants. Not only do they protect the interior and exterior of cells, but they also circulate in the bloodstream destroying free radicals before they can do damage to body components. What's more, the various nutrients in OPCs have chemical structures that protect against different types of free radicals. Whereas a single antioxidant compound, such as vitamin E or vitamin C, is protective against a number of free radicals, a mixture of many different types of antioxidants protects against a greater number of types of free radicals. Pycnogenol's unique composition is what gives it the ability to double the content of antioxidative enzymes inside the cell, in addition to directly neutralize free radicals.

Well, so much for the biochemical background of oligomeric proanthocyanidins. Hopefully this background information helps explain why OPCs are so effective, but what is more important is what the clinical studies show. The following chapters will look at the clinical studies involving OPCs' health benefits. Let's begin with heart disease.

CHAPTER TWO

Protecting the Heart and Circulation

There are several forms of heart disease and several causes for it. Yet, most people think of a heart attack as the end result of heart disease, and associate cholesterol as the only cause of it. This chapter will explore the causes of heart disease and other circulatory disorders and discuss the cardiovascular benefits of OPCs.

Early Heart Disease Research with Antioxidants

The earliest research to show that antioxidant nutrients reduced the incidence of heart disease originated from an epidemiological study I did in the mid-seventies. The study's results showing that long-term intake of vitamin E correlated with a significant reduction in the incidence of heart disease first appeared in a 1976 issue of *Prevention* magazine and were followed a year later with the publication of my book *Supernutrition for Healthy Hearts* (Dial Press, 1977). The findings revealed that moderate amounts of the antioxidant nutrient could reduce heart disease incidence by 40 percent in several age groups when taken for more than ten years.

The study of 17,894 people between the ages of fifty and ninety-eight showed a dramatic drop in heart disease among those taking vitamin E over a long period of time, averaging a 40 percent decrease. I found that the length of time vitamin E was taken was more important than the amount. This trend was especially apparent above nine years of usage. The group taking 400 IU (international units) or more of vitamin E daily for ten years or more was strongly associated with reducing the incidence of heart disease prior to eighty years of age to less than 10 percent of the normal rate. Among the 2,508 people who had been taking vitamin E over ten years, only four suffered from heart disease. Ordinarily, in a sample of that size, approximately 836 people would be expected to suffer from heart disease. Appropriate statistical models reveal that high confidence can be given to the statement that taking 400 IU of vitamin E daily for more than ten years will lower the incidence of heart disease to 10 percent of the then current (1975) age-adjusted rate of 32 per 100 to 3 per 100.

The observation that length of time taking vitamin E is strongly associated with reduced heart disease is strong and extremely significant. (The Spearman-Rho statistical association value was 0.96, where 1.00 is perfect association. The correlation is significantly different from zero or chance occurrence at the 0.01 level.)

In hindsight, the reason now seems obvious. For any given age, this means that the longer time taking vitamin E, the longer the arteries have been protected from oxidized low-density lipoprotein (LDL) cholesterol. Protection was started earlier and the arteries were cleaner.

In 1993, two epidemiological studies by Harvard University researchers confirmed my findings. Dr. Meir Stampfer and colleagues published an eight-year prospective epidemiological study of 87,245 nurses and reported that women who took more than 100 IU of vitamin E daily for more than two years had 46 percent lower risk of heart disease. Also, Dr. Eric Rimm and his Harvard colleagues published a five-year prospective epidemiological study of 51,529 male health professionals and reported that men who used more than 100 IU of vitamin E daily had a 37 percent lower risk of heart disease. Again, it should be noted that at least two years of usage is needed for a significant drop in heart disease to be observed.

Vitamin E works as an antioxidant to protect the artery linings, the proteins that transport cholesterol, and the blood platelets. I'll explain these functions shortly, but right now I want to point out that OPCs are more powerful than vitamin E in performing these functions and my opinion is that OPCs are more effective than vitamin E alone. As you may recall from Chapter 1, vitamin E is only a *mono*phenolic antioxidant whereas OPCs are *poly*phenolic antioxidants. The greater number of phenol groups in OPCs is one reason for their ability to neutralize a broad range of free radicals. It is also my opinion that the best protection against heart disease is with a synergistic combination of OPCs, vitamin E, and vitamin C.

How Free Radicals Cause Heart Disease

The path to a heart attack is a two-step process. Atherosclerosis, the medical term for narrowing of the arteries, does not by itself cause heart attacks. Thrombosis (a blood clot) and vasoconstriction (constriction and/or spasm of an artery) are the events that usually precipitate a heart attack.

First "foam cells" (white blood cells filled with oxidized LDL) adhere to the artery lining. These cells promote the infiltration of various substances through the artery wall into its middle layer. Now the artery can be said to be "diseased" as a plaque is formed in the artery interior. As the plaque expands, the wall is pushed out and the opening where the blood flows through is narrowed. Thus, blood flow is decreased to the heart tissue. The narrowed artery also damages the blood platelets passing through, making the blood "sticky" and encouraging clot formation at the plaque site.

Note that the key is that narrowed arteries (blood vessels that carry blood from the heart) put the squeeze on blood platelet cells and damage them. Platelet cells are the cells that clump and clot in the blood. If you have a cut, they clot and keep you from bleeding to death. But in blood vessels, platelet aggregation leads to clots that can interrupt the flow of blood. These clots can lodge in the narrowed arteries of the heart (coronary arteries) completely shutting off the flow of blood through that artery. When this happens, doctors call it a coronary thrombosis—a blood clot in a coronary artery. Hence, the expression that someone is having a "coronary."

When a blood clot shuts off the flow of blood in a coronary artery, the region of the heart fed by the artery is starved of oxygen and nutrients. The result is the death of these cells, which is called an infarct. This is the classic heart attack, called an acute myocardial (heart) infarction, or AMI for short.

Vasoconstriction causes reduced blood flow to the heart by constricting the diameter of the artery. It can even completely shut off an artery and stop all blood flow.

Another common form of heart disease is heart failure, in which the heart is too weak to efficiently pump blood. Usually, the heart enlarges as it tries to compensate for the reduced output. Angina is the pain experienced in the heart when there is not enough blood reaching all parts of the heart during activity. High blood pressure (hypertension) affects arteries and is a risk factor in various forms of heart disease.

Getting back to atherosclerosis, the process of forming the restrictive deposits (plaque) was once thought to be very simple. You eat food high in cholesterol and it zaps onto the arteries. Now it is known that the deposit-forming process is quite complicated and only directly related to dietary cholesterol in a minority of people. Yes, the process is very complicated, but it's important to remember that free radicals are involved and that OPCs, as antioxidants, are protective.

Cholesterol is not soluble in blood, so it is carried in particles called "lipoproteins." Two important lipoproteins are low-density lipoproteins (LDL) and high-density lipoproteins (HDL). The cholesterol carried by LDL, often called the "bad cholesterol," is carried to the cells from the liver. The cholesterol carried by HDL is often called the "good cholesterol," as it is being carried away from cells and back to the liver.

Cholesterol deposits form only when LDL becomes damaged by oxidation—that is, by free radicals. It's then called oxidized LDL. Oxidized LDL can infiltrate the artery wall and initiate a series of events that traps the cholesterol-containing oxidized LDL inside the blood vessel wall. Infection-fighting white blood cells, sensing that something is wrong, are attracted to the site and swell, forming foam cells, which then turn into the lesions commonly called cholesterol deposits.

LDL is oxidized only when the amount of antioxidants is insufficient to protect the LDL against oxidation by free radicals. Studies by Dr. A. Nelson and colleagues in 1998 showed that the OPC-containing Pycnogenol® directly protects LDL (Nelson et al., 1998). However, Pycnogenol can also indirectly protect LDL. Pycnogenol can recycle vitamin C, which, in turn, can recycle vitamin E, still the prime antioxidant that protects LDL.

Pycnogenol can also destroy free radicals before they reach the LDL and cause damage. The tendency to form oxidized LDL, and hence the cholesterol deposits (atherosclerosis), is dependent on two factors: the amount of LDL in the blood and the balance between the antioxidants and free radicals present. Both are important but the antioxidant/free-radical balance is the most important of the two.

It's also important to recognize that the so-called bad cholesterol, the LDL, is not bad unless it is deprived of antioxidants. LDL is needed to transport fat-soluble antioxidants (such as vitamin E) through the bloodstream, which means it's essential for health. But like the rest of the body, LDL cholesterol also needs antioxidants to stay healthy.

Remember, cholesterol deposits by themselves don't cause a heart attack. They are a major contributing factor to forming the blood clot that causes the heart attack. As long as the blood can squeeze through the narrowing caused by the cholesterol deposits in good volume, the heart will receive sufficient oxygen and nutrients to keep the heart tissue alive. A critical factor then is to maintain the proper "slipperiness" of the blood cells and prevent a clot from forming in the coronary arteries. OPCs have a protective anti-aggregation (anti-clotting) effect on blood platelets, and are particularly effective against the damage to platelets.

OPCs Keep Blood Flowing and Blood Vessels Relaxed

Blood platelets are small, disc-shaped, colorless blood cells that are smaller than red blood cells. There are about 150,000 to 300,000 platelets per cubic centimeter of blood. It is uncanny to me just how much a platelet looks like a chocolate-chip cookie when observed through an electron microscope. Even granular proteins on the platelet surface resemble chocolate chips.

Platelets play a major role in the process of coagulation of blood to arrest bleeding (hemostasis). When bleeding begins, the vessel constricts, a protein called "tissue factor" is released, and a protein (collagen) in the vessel wall is exposed. When tissue factor is released by the blood vessel wall, a lipoprotein on the surface of the platelet called "platelet factor 3" is activated and reacts with blood factors to promote the formation of a platelet plug and initiate other steps in the blood-clotting mechanism. When platelet factor 3 is activated, the platelet changes shape and is said to be "activated."

However, platelets can be activated even when there is no bleeding, and this is not good. If they are activated, they still tend to aggregate or clump together and initiate an undesirable blood clot that can block blood flow through the vessel, and result in a heart attack or a stroke. If this undesirable blood clot is stationary, it is called a thrombus and if it travels through the vessel it is called an embolism.

Platelets can be activated by tobacco smoke and stress, as well as by diabetes and certain nutritional deficiencies. As people age, a greater percentage of their platelets tend to be undesirably activated.

Stress produces the same undesirable activation of blood platelets that smoking, diabetes and nutritional deficiencies do. When we are under stress, our adrenaline really flows. Adrenaline activates blood platelets to be prepared to clump together and form a blood clot. Furthermore, adrenaline causes blood vessels to constrict with the consequence of higher blood pressure. These physiological changes were an advantage many years ago when we were at risk of being attacked by wild animals. Should an injury to occur, the blood platelets were prepared to more quickly form a blood clot and more efficiently prevent blood loss.

Nowadays, we live with constant stress—on the job and at home. It seems we never have time to relax. This stressful condition causes permanent constriction of blood vessels, as well as continuously sticky platelets.

As a consequence, the diameter of blood vessels is reduced, leaving less space for the blood to flow. An atherosclerotic plaque may further reduce the space for blood flow. If now some platelets suddenly stick together and form a clot, the vessel may be completely clogged and the surrounding tissue is no longer supplied with nutrients and oxygen. This is how heart infarction and stroke develop.

Fortunately, OPCs helps protect us from the damage of stress. OPCs can't make the cause of our stress go away. But important OPC-rich Pycnogenol can improve the nitric oxide levels produced by the cell linings of arteries and blood vessels (Fitzpatrick et al., 1998). The nitric oxide instructs muscles surrounding the blood vessels and causes them to relax. At the same time, nitric oxide also instructs the platelets to return back to their normal nonsticky condition, and to no longer be "alarmed." Thus Pycnogenol helps the body's own mechanism to counteract the activity of the stress hormone adrenaline. Pycnogenol helps keep the blood slippery (as an anticoagulant) to reduce the chances of heart attacks and strokes.

Studies conducted by Dr. Peter Rohdewald in Germany and Dr. Ronald Watson in the United States show that Pycnogenol blocks the effect of adrenaline on blood platelets, thereby reducing the platelets' tendency to clump together to form blood clots. When a person is under stress, large amounts of adrenaline are released, which cause the blood platelets to clump together. Pycnogenol is particularly effective against increased platelet aggregation (stickiness and increased clotting tendency) caused by smoking (Pütter et al., 1999).

Dr. Watson's research, published in *Cardiovascular Reviews Reports* in 1999, was entitled "Reduction of Cardiovascular Disease Risk Factors by French Maritime Pycnogenol" (Araghi-Niknam et al., 1999). A joint article by Drs. Watson, Rohdewald, and their colleagues was published in *Thrombosis Research* in 1999 as "Inhibition of Smoking-Induced Platelet Aggregation by Aspirin and Pycnogenol" (Pütter et al., 1999).

In the clinical studies, the protective effect of a single dose of 100 mg (milligrams) of Pycnogenol was observed within two hours, and this protective effect lasted for twelve hours after ingestion, and a dose of 200 mg protected the blood for up to two to three days. When smokers took 200 mg of Pycnogenol daily for sixty days, their blood platelets returned to almost normal for nonsmokers.

Dr. Watson has compared the tendency to form blood clots in a group of elderly people (average age sixty-five) with a group of younger people (average age thirty-two). He showed that the tendency of elderly people to form blood clots is significantly higher than in younger people—a finding that helps explain the higher incidence of thrombus formation, heart infarction, and stroke with increasing age. After daily supplementation with Pycnogenol over eight weeks, the tendency to form blood clots in the group of elderly people was almost normalized to the value found in the group of the thirty-two year olds (Araghi-Niknam, 1999). Pycnogenol improves blood circulation in the elderly through an anti-thrombotic mechanism and by causing vasodilation through optimal production of nitric oxide from the endothelium.

This protective effect of Pycnogenol is achieved by supporting the production of the body's own messenger molecule nitric oxide (somewhat like a hormone). As previously explained, nitric oxide is produced by cells forming the inner lining of blood vessels. It acts on blood platelets to "calm them down," to stop them being in a state of alarm and sticky, ready to form a blood clot.

Pycnogenol hinders platelet aggregation by inhibiting the activity of the enzymes thromboxane A2, 5-lipoxygenase, and other clotting compounds. Moreover, this protection comes without an increased risk of prolonged bleeding times, or the side effects common to aspirin.

In addition, Pycnogenol decreases the level of thromboxane A2, which is also a vasoconstrictor (a substance that constricts blood vessels, thereby reducing blood flow and increasing blood pressure) in smokers to the normal level of nonsmokers (Araghi-Niknam et al., 1999).

An important point is that this beneficial effect of Pycnogenol is different from the effect of aspirin on blood platelets. The difference, though technical, is important. OPCs support the body's own production of nitric oxide, whereas aspirin irreversibly inhibits the enzyme cyclooxygenase, which leads to undesirable bleeding problems.

Aspirin is widely prescribed by cardiologists to protect against heart attacks. Early studies showed that taking the proper dosage of aspirin can reduce the incidence of another heart attack in heart patients. Later studies showed that aspirin can also reduce the risk of having a first heart attack.

So far, this sounds good, but unfortunately, many people develop serious problems with prolonged aspirin use. They can develop ulcerated linings of the gastrointestinal tract and an increased tendency to bleed. Some people have been known to develop this condition suddenly and without warning. While aspirin therapy has benefit for many people, you should check with your doctor before taking it on a long-term basis.

On the other hand, OPCs are safe and do not cause increased bleeding or the side effects of aspirin. In the studies by Drs. Rohdewald and Watson, it was found that 100 mg of Pycnogenol achieved the same desired effect on blood platelets in smokers as 500 mg of aspirin—and without the prolonged bleeding and other side effects of aspirin (Pütter et al., 1999).

Damage to the endothelium, or lining of the heart and arteries, can also cause clots to form and allow cholesterol carriers to enter the artery walls. Researches at Loma Linda University, in Southern California, studied the protective effect of Pycnogenol using arterial endothelial cells. They found that Pycnogenol reduced the damage to endothelial cells caused by free radicals and through other noxious elements (Liu et al., 2000). They had earlier noted that Pycnogenol increased the production of other antioxidants in the cells (Wei et al., 1997).

OPCs Lower High Blood Pressure

Pycnogenol also has a mild hypotensive (blood pressure-lowering) effect that helps prevent high blood pressure. This effect is important, but it does not make Pycnogenol an antihypertensive drug. There are two known reasons for this action. One mechanism, alluded to earlier, involves the optimization of nitric oxide production in the blood vessels. Several researchers, including Dr. David Fitzpatrick of the University of South Florida and Dr. Lester Packer of the University of California, Berkeley, have conducted research on Pycnogenol and nitric oxide, and how Pycnogenol balances nitric oxide levels in the artery linings to facilitate blood flow. The second mechanism, which I'll discuss momentarily, involves OPCs' ability to inhibit an enzyme in the blood that constricts or narrows the arteries.

Nitric oxide has recently aroused much interest among scientists, although it was dismissed for decades as not being an important compound in the body—but merely a waste product or an inhaled air pollutant. Nitric oxide was named the "Molecule of the Year" in 1992 by the prestigious journal *Science*. In 1998, three scientists were awarded Nobel Prizes for their research on nitric oxide. Now, we understand that it has far-reaching effects throughout the body.

Nitric oxide plays a role in many biochemical functions. It improves memory and attention; it increases perfusion of kidneys, lungs, and liver by enhancing blood flow; it preserves the functioning of the cardiovascular system; and it is responsible for the male erection.

In terms of heart disease, readers may be familiar with the fact that during angina attacks (chest pains) patients find relief from taking nitroglycerin pills. These nitroglycerin pills release nitric oxide, which relaxes the coronary arteries and allows more blood to flow into the heart.

As previously described, the inner lining, or endothelium, of the arteries produces nitric oxide, which plays a role in the regulation of blood flow. In addition, the nitric oxide produced in the artery lining also acts to increase the production of a chemical messenger called "cyclic-GMP" (guanosine monophosphate), which is needed to keep blood platelets slippery and not prone to clumping or aggregation. Pycnogenol stimulates the enzyme nitric oxide synthase to produce nitric oxide in the artery linings from the amino acid arginine.

Some nitric oxide is always needed, but too much can kill cells. Pycnogenol helps regulate nitric oxide levels in the body at optimal levels. It helps the body produce adequate levels of nitric oxide for necessary functions, while reducing the levels of nitric oxide where it does harm.

Dr. Fitzpatrick also conducted tests to determine the effect of Pycnogenol on portions of the aorta, the main artery that carries blood from the heart. He found that it improved the production of nitric oxide in the endothelium, which in turn had a relaxing effect on the aorta in a dose-dependent manner. The results were published in 1998 in the *Journal of Cardiovascular Pharmacology* (Fitzpatrick et al., 1998).

OPCs also aid the heart by blocking some of the action of angiotensin I converting enzyme (ACE) in causing high blood pressure. In this way, OPCs are similar to, but much safer than, common prescription drugs called ACE inhibitors. ACE interferes with bradykinine, a compound that helps keep peripheral blood vessels properly dilated. Blocking this action leads toward a normalization of blood pressure without a danger of driving the blood pressure too low. It allows the bradykinine to act as it should, unencumbered by ACE.

Dr. Miklos Gabor and his colleagues at the Albert Szent-Györgyi Medical University in Szeged, Hungary, along with Dr. Peter Rohdewald of the University of Muenster, Germany, found that the important OPC, Pycnogenol, has a dose-dependent action in blocking ACE from raising blood pressure. Based on their study, published in 1996 in *Pharmaceutical and Pharmacological Letters*, the researchers described the hypotensive effect of Pycnogenol as "moderate,"—people with normal or low blood pressure will not be affected—whereas those with high blood pressure due to too much ACE will benefit.

Furthermore, a clinical study published by Dr. Ronald Watson, professor in the Department of Nutritional Sciences at the University of Arizona, in 2001, examined this possibility. Dr. Watson and his colleagues' study was a randomized, double-blind, placebo-controlled, prospective, sixteen-week crossover study to determine the role of a Pycnogenol in modifying blood pressure in mildly hypertensive patients.

They found a significant decrease in the systolic blood pressure during Pycnogenol supplementation. (Systolic pressure represents the pressure in the arteries as the heart contracts and pumps blood into the arteries.) Also, serum thromboxane concentration was significantly decreased during Pycnogenol supplementation (Araghi-Niknam et al., 1999). The scientists concluded, "Our data suggest that supplementation with Pycnogenol is effective in decreasing systolic blood pressure in hypertensive patients."

In 2004, researchers led by Dr. X. Liu confirmed that Pycnogenol improved the endothelial function of hypertensive patients and allowed those on an antihypertension drug to decrease their dosage (*Life Sciences*, 2004).

OPCs Protect against Harmful Side Effects of High Blood Pressure

Pycnogenol is also of value to people who are already being treated for high blood pressure. A study published in a 2006 issue of the *Journal of Clinical and Applied Thrombosis/Hemostasis* shows Pycnogenol reduced edema, a typical side effect of antihypertensive medications, by 36 percent in patients taking these medications.

According to Dr. Gianni Belcaro, lead researcher of the study, more than 35 percent of patients taking antihypertensive medications are believed to suffer from edema as a side effect. This happens because the antihypertensive medications cause blood vessels to dilate, which allows easier blood flow and thus lowers blood pressure. However, as a side effect, this causes blood to pool in the vessels of the lower legs. As a result, they stretch and liquid seeps into tissue causing swelling (edema). Hypertension is a serious risk factor for developing severe cardiovascular incidents some time in the future and thus the necessity for treatment justifies the development of edema as a side effect.

Antihypertensive medications reduce pressure by inhibiting constriction of blood vessels. "The larger the blood vessel diameter, the easier blood will flow with less pressure," said Dr. Belcaro. "In order to avoid blood pooling in the lower legs and feet (edema), blood vessel diameters must adjust when a person changes positions from laying down to standing up. The results of this study show Pycnogenol to improve blood circulation, avoiding blood pools, and reducing edema."

The study sampled fifty-three hypertensive patients at the G. D'Annunzio University in Italy. All patients suffered from edema of their ankles and feet as a result of antihypertensive medications and were taking medications at the same dosage for at least four months. Twenty-three patients were being treated with ACE inhibitors (brand names Mavik, Altace) and thirty patients were being treated with the calcium channel blocker nifedipine (brand names Adalat, Procardia).

In the eight-week study, twenty-seven patients were given 150 mg of Pycnogenol per day versus an equivalent dosage of a placebo (an inert pill) for the remaining twenty-six patients.

The blood vessels causing edema of their lower legs were measured using a strain gauge plethysmography (a general instrument for determining and registering variations in the size of an organ or limb, in this case, the lower leg). Patients were first measured in supine position then while standing up.

After eight weeks of Pycnogenol treatment, patients treated with ACE inhibitors experienced a 35 percent decrease of ankle swelling while patients being treated with nifedipine experienced a 36 percent decrease of ankle swelling. According to Dr. Belcaro, Pycnogenol helps defy a major side effect of antihypertensive medication.

A follow-up study confirmed that Pycnogenol protects the heart against the effects of high blood pressure. The study, published in a 2007 issue of the journal of *Cardiovascular Toxicology*, indicates that

Pycnogenol helps prevent damage that high blood pressure causes to the heart. The study demonstrates that Pycnogenol can counteract the "wearing out" of the heart, which may aid the five million Americans living with heart failure.

In hypertension, the overworked heart gradually wears out, resulting in the weakening of the heart muscle and the enlarging of one or more of the heart chambers. This process (known as cardiac remodeling) makes it increasingly difficult for the heart to pump sufficient supplies of oxygenated blood to meet the body's needs, and may eventually cause heart failure. The study showed that Pycnogenol prevents the heart from getting worn out during hypertension. Cardiac chamber walls showed a significantly higher rate of collagen connective tissues than control groups.

"Alternative treatments such as Pycnogenol are crucial components in the fight against heart disease," said Dr. Watson, a lead researcher in the follow-up study. "The effectiveness of Pycnogenol supplementation is a great option for many people who want an alternative to prescription medications such as beta blockers or ACE inhibitors. This new study shows Pycnogenol limits the degenerative process in patients predisposed to congestive heart failure, such as the aged."

The study was conducted at the University of Arizona, Tucson. Elderly female mice (eighteen months old) were randomly divided into four groups: control mice, mice receiving Pycnogenol only, mice receiving L-NAME only (NG-nitro-L-arginine methyl ester, a substance that causes arterial constriction and raises blood pressure), and mice receiving both Pycnogenol and L-NAME. Pycnogenol and L-NAME were administered in tap water and the study was approved by the Animal Review Committee at the University of Arizona.

One group of mice made hypertensive by L-NAME received Pycnogenol in drinking water for four weeks and another group of mice made hypertensive by L-NAME was left untreated. After five weeks, the hearts of the latter control group had significantly increased in size as a result of hypertension. In the Pycnogenol-treated group, hypertension and heart function parameters resembled those found in the healthy control mice with healthy blood pressure.

"This study provides evidence that oral administration of Pycnogenol reversed cardiovascular remodeling induced by L-NAME by blocking nitric oxide production, which leads to hypertension, and finally cardiomyopathy [inflammation of the heart muscle]," said Dr. Watson.

After a detailed investigation of the heart tissue, Dr. Watson found that Pycnogenol supplementation had significantly enhanced the connective collagen matrix of the cardiac tissue. Whereas the chronic hypertension in mice led to a significant loss of connective collagen fibers, Pycnogenol significantly increased the collagen presence, resulting in stronger cardiac chambers.

According to the American Heart Association nearly five million Americans are living with heart failure, and 550,000 new cases are diagnosed each year. The mortality rate for heart failure affects 80 percent of men and 70 percent of women within twelve years of obtaining cardiovascular diseases. According to Watson, cardiac remodeling is considered an important therapeutic target to battle this disease.

OPCs Quench Inflammation in the Arteries

Evidence shows that chronic inflammation from ordinary bacterial infections significantly increases the incidence of heart disease. It may seem strange to find that ordinary infections, such as periodontal (gum) disease, sinus infections, bronchitis, urinary tract infections, and stomach ulcers, are linked to heart disease, but inflammation also activates infection-fighting white blood cells, which use free radicals to destroy bacteria and other "foreign" objects in the blood. These white blood cells migrate to the arteries where some of the free radicals leak out of the cells, oxidizing LDL and damaging the linings of arteries. The process also elevates a compound called C-reactive protein (CRP) in the blood. Doctors are now considering elevated CRP levels as a risk factor in heart disease.

In a 2001 article in the American Heart Association's journal *Circulation*, it was reported that chronic infections may triple the risk of atherosclerosis in people, even when they don't have the classical risk factors such as high blood pressure, obesity, diabetes, or a sedentary lifestyle.

OPCs help in a couple of ways. As antioxidants, they neutralize the free radicals released by inflammation. Dr. Benjamin Lau and his colleagues at California's Loma Linda University showed that the OPCs in Pycnogenol decrease the body's production of cellular adhesion molecules (CAM), which causes inflammatory cells to stick to blood vessel walls (Peng et al., 2000).

I will discuss inflammation in more detail in Chapter 8 on arthritis and asthma.

OPCs Help Maintain Proper Blood Circulation

OPCs help maintain good circulation in several ways. I have already discussed how OPCs improve blood flow via their effect on nitric oxide. Another way is that OPCs protect the endothelial cells that line the heart and blood vessels against free radicals. If these cells were damaged, the body would try to repair them, and this would result in scarring and lesions that reduce the flow of blood.

OPCs also bind to collagen and elastin, which are important proteins in blood vessels and skin. While bound to these proteins, OPCs block their degradation by certain enzymes. OPCs also facilitate the production of "ground substance," an intercellular cement that can fill much of the space between cells in the blood vessels and control the amount of fluid and compounds that can slip through. This also gives the blood vessels strength.

A very interesting study was conducted in Beijing, China, and published in the *European Bulletin of Drug Research* in 1999. The study was led by Dr. Shiwen Wang of the Institute of Geriatric Cardiology at the PLA General Hospital in Beijing (Wang et al., 1999). Sixty heart patients were studied for four weeks in a double-blind, placebo-controlled, randomized clinical trial. Blood micro-circulation was measured by the diameter of the capillaries in the fingernail bed. Patients were monitored for adverse heart effects, including abnormal electrical activity (using an electrocardiogram, ECG), myocardial ischemia (reduced blood flow), and blood platelet aggregation, among other parameters.

Of the heart patients given 150 mg of the OPC-containing Pycnogenol three times a day for four weeks, 54 percent had improved microcirculation versus only 31 percent of those receiving a placebo.

The study concluded that Pycnogenol inhibited the adhesion and aggregation of platelets, enhanced the diameter of microvessels and microcirculation perfusion, and to some extent prevented and improved the myocardial ischemia in patients with coronary artery disease. There were no severe side effects or toxicity to vital organs. These findings suggest that Pycnogenol can be used as a beneficial health protection agent to help preventing cardiovascular damage and thrombotic coronary artery disease (Wang et al., 1999).

OPCs Contribute to Overall Heart Health and Maintenance

I have discussed many ways in which just one of the complex of OPCs, Pycnogenol, protects against heart disease. In 2003, Dr. Watson, whose research I discuss throughout this chapter, commented on the many ways Pycnogenol protects against heart disease in a scientific article. Dr. Watson concluded the supplement has the potential to counteract major cardiovascular risk factors. The article, published in *Evidence-Based Integrative Medicine*, provides an overview of clinical studies demonstrating Pycnogenol's heart-health benefits, including reducing platelet activity, lowering high blood pressure, relaxing artery constriction, and improving circulation.

"I have worked extensively with Pycnogenol in the last decade, studying and conducting research on this super-strength antioxidant. I find it intriguing that studies carried out by different researchers across the globe have similar results to mine, with support to the concept that Pycnogenol supplementation—between 25 mg and 200 mg [per day]—helps to combat cardiovascular risk factors and contribute to overall heart health and maintenance," said Dr. Watson.

According to the article "Pycnogenol and Cardiovascular Health," British scientists recently submitted a proposal that suggests supplementing individuals fifty-five years of age and older with a drug combination in the form of a single pill. This original "polypill" concept could reduce cardiovascular risk factors, which include lowering bad (LDL) cholesterol and blood pressure, reducing platelet activity, and improving circulation.

"While we find this 'polypill' concept appealing, it's important for us to recognize that overall cardiovascular health is attainable through nutritional methods. To date, over ten published clinical studies on Pycnogenol demonstrate its significant benefit for heart and circulatory health. The majority of the research conducted provides solid evidence that this antioxidant is powerful in reducing inflammation in the body, strengthening the vascular system, lowering high blood pressure and cholesterol, and fighting the effects of smoking, stress, and other environmental risk factors on the heart, " commented Dr. Watson.

Considering all of the synergistic components in addition to Pycnogenol, OPCs offer substantial and significant protection against the major risks of heart disease. Let's look at a couple of studies involving grape seed extract, another important complex of OPCs.

Grape Seed Extract and Heart Health

In a survey published in 1993 in the *Journal of the American College of Cardiology*, Drs. D. T. and S. D. Nash of the State University of New York Health Science Center at Syracuse and their colleagues showed in a sample group of fifty-six men and women using up to 1.5 ounces (43 grams) per day of grape seed oil—an amount that one can cook with—that grape seed oil had the ability to raise healthy HDL levels by 13 percent and reduce harmful LDL levels by 7 percent in just three weeks. The subjects' total cholesterol/HDL ratio was reduced 15.6 percent and their total LDL/HDL ratio was reduced by 15.3 percent, which could be significant for those at risk of heart attack.

Dr. J. Yamakoshi and colleagues demonstrated that grape seed extract prevented plaque-causing damage to LDL in laboratory animals (*Atherosclerosis*, 1999). The grape seed extract lowered the oxidative level of the blood by 90 percent and thus prevented oxidation of the LDL. In the study, rabbits were fed a high-cholesterol diet with and without grape seed extract. Although there was no change in cholesterol levels in the rabbits, those receiving the grape seed extract showed a reduced level of damaging cholesteryl ester hydroperoxides and aortic malondialdehyde, two substances that damage LDL and can lead to atherosclerosis. Analysis showed a decrease in oxidized LDL.

In another experiment using human plasma, grape seed extract added to the plasma also inhibited the oxidation of LDL, suggesting a possible benefit in preventing atherosclerosis.

Red Wine Extract and Heart Health

A number of epidemiological and case-control studies have shown that those individuals who consume one to two glasses of red wine per day are at reduced risk for heart disease. A third important complex of OPCs and other polyphenols is red wine extract.

Dr. M. J. Halpern and his colleagues at the Superior Institute for Health Sciences in Lisbon, Portugal, found that red wine extract reduces platelet aggregation (*Journal of International Medical Research*, 1998).

In 2000, Dr. L. Fremont and colleagues at the Laboratoire de Nutrition et Sécurité Alimentaire, in Jouy-en-Josas, France, found that red wine protected polyunsaturated fatty acids in plasma and membranes. The researchers concluded, "that in humans, red wine extract may be beneficial by reducing the deleterious effects of a dietary overload of cholesterol" (*Lipids,* 2000).

In 2002, Dr. C. Auger and colleagues at the Unité Nutrition, Laboratoire Génie Biologique et Sciences des Aliments, Université Montpellier, France, published a study in the *Journal of Nutrition* on the effects of red wine extract on plasma lipoproteins and early atherosclerosis in hamsters. Plasma cholesterol and triglyceride concentrations were lower in groups that consumed the red wine extract compared to placebo controls. There was a significiant decrease in the concentration of plasma apolipoprotein (Apo) B, a damaging protein in LDL. Activity of the antioxidant enzyme glutathione peroxidase was 67 percent greater in the group receiving red wine extract. Fatty streaks in the aorta area were significantly reduced by 32 percent in the group receiving red wine extract in comparison with their respective placebo controls. The researchers concluded, "that chronic ingestion of red wine extract prevents the development of atherosclerosis through several mechanisms."

In 2006, the red wine antioxidant resveratrol was found to be cardio-protective. In a study involving rats, administration of resveratrol was shown to have dramatic cardioprotective effects (*Vascul Pharmacol,* 2006). Results showed that rats given resveratrol before having an experimentally induced heart attack were found to have significantly better post-ischemic ventricular functional recovery, less severe of a heart attack, and less destruction of cardiac tissue, as well as increased nitric oxide and decreased malonaldehyde in the heart muscle.

In the same year, a U.S. epidemiological study suggested that Mediterranean diets rich in resveratrol are associated with a reduced risk of coronary artery disease. Researchers led by Dr. A. Csiszar at New York Medical College studied the effect of resveratrol because TNF-alpha-induced endothelial activation and vascular inflammation play a critical role in vascular aging and the formation of plaque depositis on blood vessel walls. TNF-alpha, short for tumor necrosis factor alpha, is a protein that is naturally produced by the body to mobilize white blood cells to fight infections and other invaders. If the body does not remove the TNF-alpha, it continues to build up, which can cause excessive inflammation, pain, and tissue damage throughout the body.

Researchers evaluated whether resveratrol inhibits TNF-alpha-induced signal transduction in human coronary arterial endothelial cells (HCAECs). They concluded, "We found that TNF-alpha significantly increased adhesiveness of the monocytic THP-1 cells [a cell line of human macrophages] to HCAECs, an effect that could be inhibited by pretreatment with resveratrol and pyrrolidine dithiocarbamate," a synthetic antioxidant and inhibitor of nuclear factor-kappa B (NF-kB), an immune cell "trigger" for inflammation. "Resveratrol also inhibited H_2O_2-induced monocyte adhesiveness," they noted. In TNF-alpha-treated HCAECs, resveratrol significantly attenuated expression of NF-kappa B-dependent inflammatory markers . . ." the researchers said.

"Thus resveratrol at nutritionally relevant concentrations inhibits TNF-alpha-induced NF-kappaB activation and inflammatory gene expression and attenuates monocyte adhesiveness to HCAECs," the researchers concluded. "We propose that these anti-inflammatory actions of resveratrol are responsible, at least in part, for its cardioprotective effects." Dr. Csiszar and colleagues published their study in the *American Journal of Physiology—Heart and Circulatory Physiology.*

As a result of these and other studies, there is mounting scientific evidence showing that OPCs—particularly those found in French maritime pine bark extract, grape seed extract, and red wine extract—are each protective against cardiovascular disease. When used in combination, however, they become synergistic, and thus more effective and powerful than when either is taken alone.

CHAPTER THREE

Reducing the Risk
of Cancer

———

OPCs may have important roles to play both in the prevention of cancer and as an adjunct to cancer therapy. OPCs also have beneficial effects on the immune system, and are protective against degenerative diseases, such as cancer, arthritis, and diabetes. However, these findings are not meant to suggest that OPCs are a "cure" for cancer.

In the 1960s during longevity studies with antioxidants that a research colleague and I were conducting, we discovered that the trace mineral selenium and certain combinations of antioxidant nutrients dramatically reduce the incidence of cancer. In laboratory studies in which carcinogens were fed to animals in their daily diet, the incidence of cancer in the animals was above 90 percent, whereas, in the animals receiving the antioxidant combination, the incidence of cancer was less than 10 percent. In studies in which no carcinogens were fed to the animals, the cancer incidence was essentially zero in those animals receiving the antioxidant combination, whereas the animals not receiving the antioxidant supplementation had a normal rate of cancer occurrence (*American Laboratory*, 1973).

Since then my colleague and I have received patents for the use of selenium alone, and in combination with other antioxidants, for the prevention of cancer (U.S. Patent No. 6,090,414). The National Cancer Institute (NCI) sponsored a confirming clinical study involving selenium led by Dr. Larry Clark, which was published in 1995 and who since then has been conducting additional clinical studies to further confirm this. To most of us in cancer research,

NCI's published findings confirmed our study and patent, however, the NCI will not confirm, "that selenium prevents cancer" until they understand the mechanism completely. The NCI has yet to conduct confirmation studies with our synergistic combination of antioxidant nutrients. With the role of antioxidants in cancer prevention so obvious and with the thousands of confirming laboratory animal and epidemiological studies, the NCI is negligent in not conducting these studies.

How Free Radicals Cause Cancer

Cancer is not a single disease, but actually a group of similar diseases. There are about 100 different types of cancer, and they all involve an abnormality of some of the body's cells. Cancers are merely cells growing wildly or uncontrolled. Most cancers involve tumors. The NCI explains the production of tumors as follows:

"Healthy" cells that make up the body's tissues grow, divide, and replace themselves in an orderly way. This process keeps the body in good repair. Sometimes, however, normal cells lose their ability to limit and direct their growth. They divide too rapidly and grow without order. Too much tissue is produced and tumors begin to be formed. Tumors can be either benign or malignant.

"Benign" tumors are not cancerous. They do not spread to other parts of the body and they are seldom a threat to life. Often, benign tumors can be removed by surgery, and they are not likely to return.

"Malignant" tumors are cancerous. They can invade and destroy nearby tissue and organs. Cancer cells also can spread, or metastasize, to other parts of the body, and form new tumors.

The key words in this explanation are "lose their ability to limit and direct their growth." Free radicals can destroy or disrupt natural metabolic processes within a cell. They can cause severe malfunction of a cell's regulatory mechanisms. Free radicals can cause cells to mutate and start the cancer process.

The development of cancer is not the single-step of mutation alone. The major steps in the cancer process include initiation, promotion, progression, cancer, and metastasis. Just having a cell mutation will not necessarily lead to clinical cancer. Most of the time, the immune system can destroy mutated cells before they lead to cancer. Initiation of the cancer process can be triggered by free-radical damage to the genes that store our genetic material (DNA) or by free-radical involvment in the activation of a cancer gene called an "oncogene." Chemicals called carcinogens can damage the DNA, but often these chemicals do not really harm DNA until they have been altered by a free radical.

Let's examine more closely how free radicals can initiate the cancer process.

- Free radicals damage the body's cell-replicating system, which consists of deoxyribonucleic acid (DNA). DNA contains the templates to reproduce all of the cells in the body and is responsible for making you uniquely "you." If this template becomes damaged by free radicals, the body may not be able to repair all of the damage, and the result is that when the body does make a new cell, it may be a mutated cell, which can become a benign (non-spreading) or malignant (cancerous) tumor.

- Free radicals can activate oncogenes that up-regulate the genetic expression of the so-called cancer genes. Up-regulation is a process that occurs within a cell, triggered by either an external or internal signal to the cell, which results in increased expression of one or more genes and as a result the protein(s) encoded by those genes.

- Cells regulate their proliferation by their ability to sense the population of neighboring cells. Free radicals can damage a cell's membrane and inactivate the sensory mechanisms in the membrane that limit cell growth and reproduction. If a sensor becomes damaged, then cell proliferation and growth become uncontrolled.

- Free radicals can activate carcinogens or pre-carcinogens such as tobacco smoke and other pollutants that start the chemical reactions that lead to cancer.

- Free radicals can suppress the immune system, inactivating the body's defense against cancer.

Even if several cells become mutated, cancer will not develop until these cells can reproduce more mutated cells and become associated in such a way as to develop their own blood supply and defense system. This second step is called promotion. Some factors, such as alcohol consumption and cigarette smoking, can speed this process. However, the immune system can still stop progression of the mutated cells at this stage until the cells develop their defenses against the body's immune system.

Antioxidants can stop or slow each of the steps in cancer development. Preliminary evidence suggests that antioxidants can also reduce the chances of metastasis and boost the immune system. A very healthy immune system can even destroy full-blown clinical cancers.

The Cancer-Preventing Ability of OPCs

Many epidemiological studies have affirmed that diets rich in fruits and vegetables reduce the incidence of several cancers. Many scientists believe that the reason fruits and vegetables are so protective is that they are rich in antioxidants, especially vitamin C and bioflavonoids.

OPCs may help protect against the very early causes of cancer by:

- Destroying cancer-causing free radicals;

- Inactivating carcinogens;

- Preventing activation of oncogenes;

- Boosting the body's immune system so that any mutated cells can be destroyed before they become cancerous; and

- Reducing the tendency of cancer cells to stick together and adhere to other sites, and thereby metastasize.

In addition, I believe that OPCs will also be found to inhibit several tumor promoters. This effect has been demonstrated with other bioflavonoids and explains part of their protective actions against cancers.

Strong oxidants and oxygen free radicals, together known as reactive oxygen species (ROS), can cause a complex in cells called nuclear factor-kappa B (NF-kB) to dissociate and enter the cell nucleus where it binds to DNA and activates oncogenes. Dr. Benjamin Lau and his colleagues at Loma Linda University in California have discovered that the OPCs found in Pycnogenol® protect NF-kB from this dissociation and from activating the oncogenes (Peng et al., 2000). Additionally, Dr. A. Nelson and his colleagues have shown that Pycnogenol can protect against damage to DNA caused by the hydroxyl radical, the most reactive of the reactive oxygen species (Nelson et al., 1998).

Pycnogenol also inhibits the formation of reactive metabolites that are produced from NNK, a nitrosamine present in tobacco that has mutagenic and carcinogenic properties (Huynh and Teel, 1998).

Proanthocyanidin-rich extracts of berries have been shown to inhibit the growth and spread of cells associated with a wide range of cancers. In a study led by Dr. Navinda Seeram from the University of California at Los Angeles's (UCLA) Center for Human Nutrition, researchers evaluated the phenolic content of six berries (red and black raspberries, cranberries, blueberries, and strawberries) and tested their ability to inhibit the growth of human cancer cell lines. Dr. Seeram reported, "Our studies provide preliminary data as to the ability of these compounds to inhibit the growth and induce apoptosis [self-destruction of a cell intended to prevent replication errors] of different human cancer cells lines in vitro" in the *Journal of Agriculture and Food Chemistry.*

The main phenolic constituents were found to be anthocyanins, ellagitannins, flavanols, flavonols, galltannins, phenolic acids, and *proanthocyanidins.* Each berry had a different and unique phenolic content.

Dr. Seeram and colleagues tested the extracts from each berry for their anti-cancer potential on a range of human cancer cell lines.

The extracts were tested for their ability to stop the spread and induction of apoptosis of human oral, prostate, breast, and colon cancer cell lines.

"It is noteworthy that the test concentrations of the berry extracts used in these cell culture experiments far exceed levels of phenolics and/or their metabolites achievable physiologically, based on current knowledge of polyphenol bioavailability," said the researchers.

Furthermore, the researchers observed that cancer cell proliferation was reduced in a dose-dependent manner. "Induction of apoptosis or cell cycle arrest can be an excellent approach to inhibit the promotion and progression of carcinogenesis [the creation of cancer] and to remove genetically damaged, preinitiated, or neoplastic cells from the body," explained Seeram.

While many different forms of phenolics are present in the berry extracts, the research suggests that the anthocyanins may be the major contributors toward programmed cell death, but significant further study is required (*Journal of Agriculture and Food Chemistry*, 2006).

Pycnogenol® boosts immunity in several ways. Dr. Ronald Watson of the University of Arizona, Tucson, specializes in studying the immune system and has conducted several studies with Pycnogenol and the immune system. In one study, Dr. Watson found that Pycnogenol boosted the levels of immune components called cytokines, which play a crucial role in organizing the tactics of an immune response toward an infection (Cheshier et al., 1996).

Dr. Lau, mentioned earlier for his research on NF-kB, showed that Pycnogenol can counteract the weakening power of the immune system by enhancing production of immune cells. This is especially important as a person ages and correlates well with the increasing risk for cancer with age.

All types of immune cells are important in the body's resistance to cancer. In an experiment by Dr. Watson, Pycnogenol partially restored the decreased levels of certain cytokines in laboratory animals infected with a retrovirus very similar to HIV (the virus that causes AIDS). In addition, Pycnogenol greatly increased the activity of another type of immune cell called natural killer cells, in infected mice. Pycnogenol also strengthens immunity by protecting the existing immune components from their own chemicals. Infection-fighting white blood cells use free radicals to destroy bacteria. When white blood cells overproduce free radicals, the cells start to commit suicide. Pycnogenol allows the bacteria to be killed while standing by to protect the white blood cells against any excess of free radicals (Cheshier et al., 1996; Liu et al., 1998; Peng et al., 1999).

On March 25, 2007 at a meeting of the American Chemical Society, it was reported that OPCs can prevent skin cancer by boosting the immune system. Proanthocyanidins show promise in animal studies as a way to prevent sunlight-induced skin cancer when used as a dietary supplement, according to researchers at the University of Alabama in Birmingham. In studies using mouse models of ultra-violet-light-induced (non-melanoma) skin cancer, mice that were fed diets supplemented with proanthocyanidins, showed a reduction in tumor number (up to 65 percent fewer) and size (up to 78 percent smaller) in comparison to control animals that did not receive the compounds, the researchers said. The compounds appear to work by inhibiting suppression of the immune system caused by ultraviolet light exposure, says Dr. Santosh Katiyar, an associate professor in the university's Department of Dermatology (section and abstract number of presentation: AGFD 011, Sunday, March 25, 10:25 A.M.).

Now let's examine what is known so far about the cancer-preventing ability of other important OPCs.

Grape Seed Extract versus Cancer

Most of the studies with grape seed extract have been confined to laboratory studies. *In vitro* (cell culture) laboratory studies have found that grape seed extract prevents the growth of breast, stomach, and lung cancer cells, but that the growth and viability of normal cells is maintained.

In 1999, Dr. X. Ye and Creighton University School of Pharmacy colleagues studied the effect of grape seed extract against human breast cancer cells, human lung cancer cells, and human gastric adenocarcinoma cells (*Molecular and Cellular Biochemistry*, 1999). In addition, they compared the effects on normal human gastric mucosal cells and normal murine (rodent) macrophage cells with the effects on the cancer cell lines. Incubation with the grape seed extract resulted in 47 percent inhibition in cancer cell growth. The grape seed extract enhanced the growth and viability of the normal human gastric mucosal cells and murine macrophage cells. The researchers concluded, "These data demonstrate that grape seed extract exhibited cytotoxicity toward some cancer cells, while enhancing the growth and viability of the normal cells which were examined."

In 2001, Drs. K. W. Singletary and B. Meline of the University of Illinois determined that grape seed extract reduced the cancer incidence of laboratory animals by 88 percent (*Nutrition and Cancer*, 2001). This closely matches what my colleague and I reported in 1973 using a combination of antioxidants. The researchers reported that "Feeding female rats diets containing 0.1–1.0% grape seed proanthocyanidins was associated with a significant 72–88% inhibition of AOM [azoxymethane]-induced aberrant crypt foci formation." AOM is a potent carcinogen used to induce colon cancer in animals. They concluded, "These results indicate that grape polyphenolics warrant further evaluation as potential colon cancer chemopreventive agents."

Grape seed extract was also effective in preventing damage to human liver cells caused by chemotherapy medications. Another cell study found that grape seed extract can prevent the cell damage, DNA damage, and cell death that occur when cells are exposed to tobacco.

Red Wine Extract versus Cancer Cells

In 2005 and also in earlier studies, Dr. G. Caderni and colleagues at the University of Florence, Italy, found that red wine extract protected laboratory animals against colon cancer induced by carcinogens. In their 2005 report they commented, "... red wine polyphenols administered with the diet to rats for 16 weeks inhibited colon carcinogenesis induced by azoxymethane or dimethylhydrazine. Polyphenol-treated animals had a consistently lower tumor yield compared to controls." They also noted, that "wine polyphenols administered to rats not treated with carcinogens, produced a significant decrease in the basal level of DNA oxidative damage of the colon mucosa (reduced by 62%)" (*Mutation Research*, 2005).

The researchers further explored the effects of red wine extract by studying gene expression profiles. Interestingly, they observed that two major regulatory pathways were down regulated in the colon mucosa of red wine extract-treated rats: inflammatory response and steroid metabolism. (Down-regulation is a process resulting in decreased gene and corresponding protein expression.) They also found a down-regulation of many genes that regulate cell surface antigens, metabolic enzymes, and cellular response to oxidative stress. The researchers concluded, "Reduction of oxidative damage, modulation of colonic flora, and variation in gene expression may all concur in the modulation of intestinal function and carcinogenesis by red wine polyphenols."

In 2000, a team of researchers led by Dr. L. Giovannelli at the University of Florence investigated the effect of red wine extract on DNA damage (*European Journal of Nutrition*, 2000). They examined the effects of complex polyphenols and tannins from red wine and tea on DNA oxidative damage in the colon mucosa of rats. The results showed that wine polyphenols and tannins induced a significant decrease (62 percent for pyrimidine and 57 percent for purine oxidation) in basal DNA oxidative damage in colon mucosal cells without affecting the basal level of single-strand breaks. On the other hand, tea polyphenols did not affect either strand breaks or pyrimidine oxidation in colon mucosal cells. The researchers concluded, "Dietary polyphenols can modulate *in vivo* oxidative damage in the gastrointestinal tract of rodents. These data support the hypothesis that dietary polyphenols might have both a protective and a therapeutic potential in oxidative damage-related pathologies."

Earlier in 1998, Dr. H. Kamei and colleagues at the Department of Surgery, Aichi-Gakuin University Hospital, Nagoya, Japan, had reported that when colon cancer cells and gastric cancer cells were cultured with red wine extract, the cancer cell growth was significantly suppressed (*Cancer Biotherapy and Radiopharmaceuticals*, 1998).

Another antioxidant in red wine extract, resveratrol, is under extensive study as a cancer preventive. Resveratrol has been shown to interfere with all three stages of carcinogenesis—initiation, promotion, and progression.

Bilberry against Cancer Mechanisms

The University of Illinois research group, this time led by Dr. J. Bomser, also showed that crushed bilberry increased the enzymes that detoxify carcinogens and decreased the levels of the enzymes that promote tumor growth (*Planta Medica*, 1996). The OPCs in bilberry appear to be responsible for the decrease in tumor growth. The researchers concluded, "Thus, components of the hexane/chloroform fraction of bilberry and of the proanthocyanidin fraction of low-bush blueberry, cranberry, and lingonberry exhibit potential anticarcinogenic activity as evaluated by *in vitro* screening tests."

As you can see, there is a growing body of scientific evidence that the OPCs are effective in reducing the risk of cancer. They work by way of several pathways, but mainly through their strong antioxidant effect in protecting body components, such as DNA against free-radical damage, and also by controlling the genes that express cancer.

CHAPTER FOUR

Slowing the
Aging Process

F ree radicals are hazardous molecules that age us and cause diseases. OPCs destroy free radicals, and by doing so, can slow the aging process. My research in the 1960s showed that certain combinations of antioxidant nutrients extended the average lifespans of laboratory animals by 20 to 30 percent and their maximum lifespans by 5 to 10 percent (*American Laboratory*, 1971). It was during these same studies that I discovered that the trace mineral selenium and certain combinations of antioxidant nutrients dramatically reduce the incidence of various cancers.

At that time OPCs were not available to study. Since the antioxidants in OPCs are more powerful than the substances I was using, I would like to repeat those studies today. However, others are starting studies with OPCs. Several studies have been done with the powerful OPC complex, Pycnogenol®, but there is more to aging than lifespan. I'll discuss these studies shortly.

The goal is to increase the quality of life, as well as the length of life. We want to add more years to our lives, but we also want to add more life to our years. Slowing the aging process is all about living better and longer.

How Free Radicals Cause Aging

There is no one physical or mental condition that can be directly attributed simply to the passage of time. It is not the passage of time that ages us, it is the accumulation of deleterious chemical events that deteriorates our bodies into the condition we call aging. Some of the alleged diseases or disorders associated with aging can also occur in the young. Children can have cancer, high blood pressure, arthritis, and the like, so it is not simply the number of years. What then is aging?

Aging is the process that reduces the number of healthy cells in the body. The most striking factor in the aging process is the body's loss of reserve due to the decreasing number of cells in each organ. For example, fasting blood glucose (blood sugar) levels remain fairly constant throughout life, but the glucose tolerance measurement shows a loss of response in aging. Glucose tolerance measures the reserve capacity of this system to respond to the stress of the glucose load used to challenge the system in the test. The same diminishment holds true for the recovery mechanisms of other systems. Simply stated, the aging process is the body's loss of ability to respond to a challenge to its status quo (homeostasis). The mass of healthy active cells in each organ declines as a person ages, thus the organ's function is diminished.

Now the question becomes, what causes this loss of reserve? Free-radical reactions result in the body's loss of active cells. The cumulative effect of billions of cellular free-radical reactions is the loss of cells. This happens in several ways.

1. Free-radical damage to the cell membranes can impair the cell's ability to transport nutrients into the cell and the cell dies without replacing itself.

2. Free-radical damage to cell membranes can impair the cell's ability to transport waste products out of the cell, thus the cell can strangle in its own waste. The result is that the cell dies without replacing itself.

3. Free radicals can damage the cell's DNA so that instead of the cell being replaced by another healthy daughter cell, the cell is replaced with a mutant that doesn't function correctly.

4. Free radicals can damage the lysosomal sac (membrane) and release deadly lysosomes, which are enzymes that destroy other

cell components. This leaves the cell devoid of working parts and it cannot be replaced. The cell becomes a clinker and the body becomes one cell older.

5. Free radicals can fuse proteins together in such a way that the proteins no longer function properly. This can damage the cell so that it does not perform and does not reproduce a healthy replacement.

6. Free-radical reactions form byproducts such as the age pigment lipofuscin or advanced glycosolated end products. These residues accumulate over time and interfere with cell function.

The result of many of the free-radical reactions is that the number of active cells in the body decreases. This is analogous to the light bulbs on an old theater marquee that burn out one by one. For a while, the message can still be read, but as the number of burned-out bulbs increases, eventually the message is not discernible. In the body, the cells in each organ decline but the organ still functions until a point.

The Anti-Aging Benefits of OPCs

In addition to the many health benefits of OPCs, there are important "vanity" benefits as well. One OPC complex, Pycnogenol, is often called the "skin vitamin" or the "oral cosmetic" because it rebuilds skin tissues, making them more flexible and smoother, which makes skin appear younger and healthier. It can't undo deep wrinkles or repair permanent sun damage, but it can make skin healthier and smoother.

OPCs can help improve the quality of life in many ways, as well as help reduce the risk of killer diseases.

It is not practical to study humans in scientific lifespan studies because these studies would take too long and there are too many variables. Scientists find it more practical to study laboratory animals that have much shorter lifespans. Often, the studies begin with *drosophila,* a fruit fly, and then, if these studies are promising, laboratory rodents are studied.

Lifespan studies with OPCs are beginning. Dr. L. Shuguang and his Chinese colleagues found that Pycnogenol increased the average lifespan of drosophila by 13 percent (*European Bulletin of Drug Research*, 2003). This would equate to an increase of nearly eight years for a person who otherwise would have a seventy-year lifespan. It's a long way from fruit flies to humans, but at least the study shows that more studies are justified.

Pycnogenol most likely increases lifespan by indirectly increasing cellular levels of the antioxidants superoxide dismutase and glutathione, as well as directly quenching free radicals throughout the body.

There are many diseases associated with aging. Alzheimer's disease is increasing at an alarming rate. Several studies are underway with various antioxidant nutrients. It's too early to tell the outcome of these studies, but preliminary signs are encouraging. Researchers know that Pycnogenol shields brain cells from the oxidative stress of free radicals. One of the characteristics of Alzheimer's disease is the accumulation of beta-amyloid protein. Dr. David Schubert and his colleagues at the Salk Institute of Biological Studies, in San Diego, have found that Pycnogenol prevents the toxic action of that protein against brain cells in laboratory experiments (Liu et al., 1998, 1999).

Dr. Lester Packer and colleagues at the University of California at Berkeley have found that Pycnogenol protects brain cells from damage from excessive amounts of glutamate. Glutamate is vital to the brain in small amounts, but it can be toxic at high concentrations (Kobayashi et al., 2000).

Aging also decreases memory skills in many people. Animal learning and memory studies are a good indication of what will apply to humans. Dr. Benjamin Lau and his colleagues at Loma Linda University in California investigated memory retention and learning ability of mice. They discovered that older mice given Pycnogenol for two months almost retained the mental levels of young mice (Liu, et al., 1999).

Red Wine Extract and Aging

In addition to the OPCs in red wine, another antioxidant called resveratrol has entered prominently in anti-aging research in the past few years. In 2006, one particular study which was published in the highly respected scientific journal *Nature* drew wide attention and was hailed as "the breakthrough of the year." The research was sponsored by the National Institutes of Health (NIH) and the lead researcher was Dr. David Sinclair of Harvard Medical School. In this study, mice receiving resveratrol and fed a high-fat diet lived 31 percent longer and maintained the same quality of life as animals fed a low-fat diet. In fact, resveratrol reversed 144 of 153 genes that are switched on when high-fat diets are consumed.

Dr. Sinclair stated, "The 'healthspan' benefits we saw in the obese mice [fed] resveratrol, such as increased insulin sensitivity, decreased glucose levels, [and] healthier heart and liver tissues, are positive clinical indicators and may mean we can stave off in humans age-related diseases such as type 2 diabetes, heart disease, and cancer, but only time and more research will tell."

The study, published in the November 1, 2006 issue, looked at the effects of feeding middle-aged mice (fifty-two weeks old) one of three diets: a standard mouse diet, a high-calorie (fat) diet, and a high-calorie (fat) diet supplemented with resveratrol.

At sixty weeks of age, the researchers reported that the survival curves of the high-calorie and the high-calorie/resveratrol groups began to diverge, with the resveratrol group showing a three to four month advantage in survival.

When the mice reached "old age" (114 weeks), co-researcher Dr. Joseph Baur reported that more than 50 percent of the mice fed the high-calorie diet had died compared to less than 33 percent of the mice receiving the high-calorie diet with resveratrol.

Baur reported that mice on the high-calorie diet without the resveratrol supplement were found to have increased plasma levels of insulin, glucose, and insulin-like growth factor (IGF) 1—markers that in humans predict the onset of diabetes—when compared with their overweight counterparts supplemented with resveratrol.

Pathological studies of the heart tissues of mice from the three diet groups showed that the abundance of fatty lesions, degeneration, and inflammation were significantly less for the standard diet and resveratrol-supplemented group (1.6 and 1.2 points on a relative scale of 0 to 4), compared to the high-calorie diet group (3.2 points).

"After six months, resveratrol essentially prevented most of the negative effects of the high-calorie diet," said Dr. Rafael de Cabo, of NIH's National Institute on Aging.

"This study shows that an orally available small molecule at doses achievable in humans can safely reduce many of the negative consequences of excess caloric intake, with an overall improvement in health and survival," concluded the researchers.

Earlier studies with the red wine antioxidant resveratrol centered on extending the lifespans of more primitive life forms. The mechanisms of resveratrol's apparent effects on life extension are not fully understood, but they appear to mimic several of the biochemical effects of calorie restriction. This seems to function by means of lipase inhibition, which reduces the absorption of fat through intestinal walls.

In 2003, Dr. Sinclair had published (also in *Nature*) that resveratrol significantly extends the lifespan of the yeast *Saccharomyces cerevisiae*. Later studies showed that resveratrol prolongs the lifespan of the worm *Caenorhabditis elegans* and the fruit fly *Drosophila melanogaster*. In 2006, resveratrol was also shown to extend the maximum lifespan of a short-lived fish, *Nothobranchius furzeri*, by 59 percent, and extended its median lifespan by 56 percent. Also noted in the fish, were an increase in swimming performance, an increase in cognitive performance (learning tasks), and a lack of neurofibrillary degeneration, or tangles (found in a control group). The researchers observed that "resveratrol's supplementation with food extends vertebrate lifespan and delays motor and cognitive age-related decline, and could be of high relevance for the prevention of aging-related diseases in the human population."

The evidence has strengthened over the past forty years that free radicals accelerate the aging process and are at least a secondary factor, if not the prime factor, in aging. By reducing free-radical damage with nutrient antioxidant combinations including OPCs, you can expect to add fifteen years of health to your lifespan. Antioxidant nutrients are especially effective in counteracting the environmental factors that accelerate aging. Remember, the prime goal is not an extended lifespan but a healthy lifespan. OPCs reduce many of the illnesses and disorders associated with aging, as well as the signs of aging such as skin wrinkling. In the next chapter we will look at the role of OPCs on keeping our skin healthy and younger looking.

CHAPTER FIVE

Keeping Skin Healthy and Younger Looking

OPCs are extremely important to skin health and appearance. As mentioned earlier, one OPC complex, Pycnogenol® has been called "the skin vitamin" and "the oral cosmetic" by many users. However, as explained earlier, OPCs are not true vitamins. OPCs help rebuild skin tissues, making skin more flexible and smoother, which makes skin appear younger and healthier. OPCs can't undo deep wrinkles or repair the permanent sun damage of actinic keratosis (a precancerous skin condition), but they can make skin healthier, livelier, and smoother.

OPCs improve the permeability of blood capillaries by improving the production of ground substance (an intercellular cement) between the cells and by promoting and protecting the two major skin proteins, collagen and elastin.

In addition to their protective benefits against free radicals that prevent skin damage, OPCs bind to collagen and elastin and protect these proteins from various enzymes that break them down. This action reduces the thinning of skin that develops with aging. OPCs help the skin rebuild its thickness and elasticity. Skin fullness and elasticity are essential for skin smoothness.

How Free Radicals Age the Skin

As skin ages, it loses its flexibility. This is primarily due to the cumulative effects of sun exposure, which alter the skin structure and reduce the amount of skin protein produced. You can easily see this effect. When young skin is pinched and pulled up, it will spring back quickly. When older skin is similarly pinched, it returns to position very slowly. Try this on the back of the hands of people of various ages. How does your skin do?

Have you ever noticed that the skin on the back of the necks of farmers and fishermen is thick, leathery, and deeply wrinkled compared to the skin of office workers? You can also compare the apparent age of skin on different areas of your own body. We tend to think of the skin of our bodies as being the same age as our chronological age, but the fact is that some cells are much newer than others.

Compare the smoothness of the skin on a sun-exposed area, such as the back of your neck, to the skin on a sun-protected area, such as your buttocks. The sun is what made the difference, by causing free radicals that fused molecules of the skin protein collagen together.

The proteins in young skin freely slide over each other and spring back to their normal length when stretched. As time goes by and exposure to the sun accumulates, the ultraviolet energy in sunlight interacts with fats in the skin to produce free radicals. These free radicals damage proteins in the skin and can link the proteins together. These damaged proteins do not easily slide over one another and do not recoil rapidly. This is where the strong protective effects of OPCs play their role of destroying free radicals, thereby lessening sun damage to skin.

Additional Skin-Protecting Benefits of OPCs

This protective effect can even extend to directly protecting against sunburn. People who are well protected by antioxidants find that their skin does not "burn" as quickly in the sun. Sunburn is an inflammation resulting from the free radicals produced by the effect of sunlight on fats in the skin. Free-radical damage can be limited by the scavenging effect of the antioxidants. Studies have shown that the time of exposure required for sunburn to develop can be increased with OPCs, but OPCs should not be your only protection from the sun. The use of sunblock, wearing a hat, and an awareness of exposure time are also important.

Remember that tobacco smoking constricts the blood vessels of the skin and leads to premature wrinkling. OPCs, on the other hand, facilitate blood flow in the skin and keep the skin well nourished. Yet, there is more that OPCs do for improving appearance. Let's look beneath the skin.

CHAPTER SIX

Improving Venous Health

I n addition to its protection against many diseases, OPCs also help protect against varicose veins and bruises and help reduce the severity of minor injuries. For more than sixty years, bioflavonoids have been known for their ability to improve venous health. OPCs have been shown to improve venous health, help repair some varicose veins, and reduce the occurrence of new varicose veins. Fifteen clinical studies in venous disorders have been conducted with the key OPC-complex Pycnogenol® alone, involving chronic venous insufficiency, varicose veins, thrombophlebitis, post-thrombotic syndrome, and venous stasis edema. These studies have been reviewed by Dr. Om Gulati in the *European Bulletin of Drug Research* (Gulati, 1999).

OPCs Help to Maintain Health of Capillaries

One of the earliest discoveries about OPCs was their ability to strengthen capillaries, the body's tiniest blood vessels. Early research focused on the role of Pycnogenol as either an independent factor or a co-factor with vitamin C in the maintenance of capillary health. Dr. Miklos Gabor of Albert Szent-Györgyi Medical University in Hungary conducted many studies over the years that demonstrate that Pycnogenol improves capillary permeability, and decreases capillary leakage and microbleeding.

Capillaries are not designed to be sealed against leakage. These blood vessels are the interface between the bloodstream and oxygen, nutrients, and waste products. Capillaries must be permeable enough to allow fluids to seep out of the capillaries, mix with the fluid that surrounds all of the cells, and then reenter the capillaries. If the capillaries are too permeable, too much fluid and protein seep out, resulting in edema (swelling), and even red blood cells may seep out causing bruising and red spots (petechiae), or even hematomas.

Dr. Gabor measured the leakage of fluid through capillaries by using a device he invented called a petechiometer, which applies a vacuum over a small area of skin. The strength of the vacuum can be varied. The greater the vacuum required to produce petechiae, the higher is the capillary strength. Dr. Gabor and his colleagues have found that Pycnogenol improved capillary strength within two hours and maintained it longer than eight hours. They described their findings in the journal *Phlebologie* in 1993.

OPCs Improve Symptoms of Chronic Venous Insufficiency

Dr. F. Feine-Haake, studied the benefit of 30 mg (milligrams) of Pycnogenol given three times a day (a total of 90 mg) on 100 people with varicose veins and other symptoms of chronic venous insufficiency (CVI). Eighty percent showed a clear improvement, and 90 percent found that their nocturnal leg cramps also disappeared.

Pycnogenol helps keep all of the blood vessels healthy and reduces edema in the legs, which contributes to the development of varicose veins. In double-blind, placebo-controlled studies, it has been shown that Pycnogenol greatly improves symptoms of CVI (Arcangeli, 2000; Petrassi et al., 2000).

Clinical studies and laboratory animal studies show that Pycnogenol reduces water retention and swelling in the legs due to edema, by strengthening capillaries and preventing leakage of fluids (Gabor et al., 1993; Blazso et al., 1994; 1997; Arcangeli, 2000; Petrassi et al., 2000).

Findings published in July 2006 in the journal of *Clinical and Applied Thrombosis/Hematosis* show a significant symptom reduction of CVI in patients after supplementing with Pycnogenol. Results from this study showed Pycnogenol to be more effective in reducing edema, tight calves, skin alterations, pain during walking, and swelling limbs than Daflon, a combination of the bioflavonoids diosmin and hesperidin, and a commonly prescribed drug used to treat CVI.

About 500,000 people in the United States develop leg ulcers due to CVI. If left untreated, leg and ankle swelling can lead to dangerous conditions such as deep vein thrombosis (DVT, see next section for more on this condition). Previous studies have shown Pycnogenol to be effective in encouraging improved circulation and helping to prevent travel-related DVT. Like varicose veins, spider veins also develop if edema is left untreated.

"Chronic venous insufficiency is caused when leg veins cannot pump enough blood back into the heart. When people are not active, blood pools in their leg veins, legs and ankles can become swollen," said Dr. Peter Rohdewald, a researcher of the study. "Eventually, some of the valves cannot hold the weight of excessive blood, which then adds more pressure onto the next valve further downwards. Ultimately, the inability to prevent the liquid in the blood from seeping into the tissue is what causes edema, a common condition of CVI."

Researchers at L'Aquila University in Italy conducted a comparative analysis by supplementing eighty-six patients with severe CVI with either Pycnogenol or Daflon. Each group supplemented daily for eight weeks. Patients who supplemented with Pycnogenol received either 150 mg or 300 mg, while Daflon patients supplemented with 1,000 mg.

Ankle swelling was measured before 10 A.M. to avoid the swelling effect of standing and again after thirty minutes of resting with feet elevated. Measurements were taken at the beginning of the study and after four and eight weeks of treatment. A composite, analogue score based on signs and symptoms (edema, pain, restless limbs, subjective swelling, and skin alterations/redness) was recorded by patients. A second evaluation of edema was made by another physician.

After eight weeks of treatment, patients who supplemented

with Pycnogenol experienced decreased ankle swelling by 35 percent, while Daflon treatment decreased ankle swelling by 19 percent. A composite score for edema including pain, restless legs, feeling of heavy swollen legs, and skin alterations was found to be decreased with Pycnogenol by 64 percent, whereas Daflon was only half as effective, lowering the composite edema score by 32 percent. The concentration of oxygen and carbon dioxide (CO_2) just beneath the skin in the lower legs was estimated with small sensors attached to the skin. Pycnogenol treatment was found to significantly increase tissue oxygen and lower CO_2, suggesting a considerable improvement in blood circulation to the legs. Daflon, in contrast, did not yield any significant effect on tissue oxygenation and apparently does not improve blood circulation to the legs.

"Interestingly, this study demonstrated that supplementation with a very high dosage of 300 mg Pycnogenol a day did not yield significantly better effects than treatment with 150 mg [of] Pycnogenol, with the exception of the composite edema score, which improved better with the higher dosage," said Dr. Rohdewald.

Continuous stretching from CVI permanently enlarges veins. Past studies have shown that treating edema with Pycnogenol prevents the development of spider veins. Pycnogenol also helps prevent existing spider veins from getting larger and more prominent. When edema is successfully treated, the increased pressure on veins gets normalized, preventing the veins from further increasing in size.

"Pycnogenol has demonstrated its efficacy and safety in several clinical studies, and symptoms of CVI have been reduced significantly by Pycnogenol in controlled studies. We were pleased to see that not only did Pycnogenol decrease CVI symptoms, but [also] the results were significantly more successful than the prescription drug used for treating CVI," said Dr. Rohdewald.

OPCs help maintain capillary strength and proper capillary permeability. A bruise is pooled blood beneath the surface of the skin. If the capillaries leak too much, fluid and proteins can leak through the capillaries into the neighboring spaces between cells, thereby altering the normal osmotic pressure. Eventually, even red blood cells can leak through and spontaneously cause a bruise without a direct injury to the capillaries. OPCs restore proper permeability and reduce the incidence of spontaneous bruising. With stronger capillaries, it will take a more severe injury to damage the veins and capillaries enough to allow microbleeding into the tissues.

OPCs Protect against Venous Clots

Since OPCs improve blood circulation and keep blood flowing, it is helpful for protection against swollen ankles and venous clots on long-distance air flights—the so-called economy-class syndrome.

Air travelers should consider taking OPCs prior to their flight and again mid-flight if the flight is longer than four hours. I have heard of too many people developing clots on flights, partly the result of dehydration and traveling in cramped conditions, and more having embolisms lodging in their lungs a day or two after their flight as the clot dislodges from the leg vein and travels through the body.

Economy-class syndrome is not limited to economy-class flights. This condition has been blamed for at least thirty deaths in three years at just one hospital in London, England. Sitting still in airplane seats encourages blood to pool in the ankles. The edge of the seat tends to reduce the return flow of the blood through the leg veins. One result is that the ankles can swell, but a worse result is that a blood clot can form in the deep veins of the legs (known as deep vein thrombosis, or DVT) and then travel as an embolism to the lung and cause a deadly pulmonary embolism. In addition, the reduced oxygen levels and increased radiation levels at high-altitude flights increase the need for the powerful antioxidant protection of OPCs.

When you are trapped in a plane at high altitude, breathing germs from your neighbors for hours at a time, getting zapped by cosmic radiation, and breathing low levels of oxygen, you definitely would benefit from extra OPCs. The first thing you will notice is that your shoes fit upon arrival. Airlines should distribute OPC supplements in-flight just like they used to pass out chewing gum to help ease ear pressure changes.

CHAPTER SEVEN

Preserving Eye Health

It's amazing how many people have been introduced to the benefits of nutritional supplements because they have developed eye problems. There will be many more as the baby boomers age. Dry eyes, glaucoma, and cataracts start to become a problem in the fifty-plus years. Later on, age-related macular degeneration looms as a real danger.

The eyes are particularly vulernable to free-radical damage as they are a source of entry of high-energy ultraviolet energy from the sun and have minimal blood circulation. Plus, many people have been raised to fear egg yolks, which are a rich source of the antioxidants lutein and zeaxanthin that the macula needs for protection.

The good news is that the OPC-complex Pycnogenol® has been shown to improve visual acuity, reduce cataracts, prevent diabetic retinopathy, and suspend the deterioration of retinal function that can lead to blindness.

Also, Pycnogenol's antioxidant power can help spare the dietary sparse vision-related carotenoids lutein and zeaxanthin, which protect the center of vision, the eye's macula, thus reducing the risk of age-related macular degeneration.

One recent poll listed the possibility of lost vision as the number-two fear of senior citizens, second only to the fear of cancer. OPCs may be important to the body's defense against both.

"Eye-Opening" Study Results Using OPCs

A double-blind, placebo-controlled study published in *Phytotherapy*

Research in May 2001, demonstrated that Pycnogenol improved visual acuity in patients suffering from diabetes, atherosclerosis, and other diseases (Spadea and Balestrazzi, 2001).

The parameters used to measure eye health included the Snellen eye chart (a measure of visual acuity), visual field, ophthalmoscopy, fluoroangiography, and electroretinography. Those receiving 50 mg (milligrams) of Pycnogenol, three times a day, had either an improvement in visual acuity or a slower rate of loss of acuity compared to those in the control group receiving a placebo rather than Pycnogenol.

Five clinical studies conducted in Europe show that Pycnogenol greatly improved symptoms in patients with diabetic and vascular retinopathies, maculopathies, and other visual dysfunctions. The vision of treated patients not only stopped decreasing further, but even improved (Spadea and Balestrazzi, 2001). Pycnogenol can improve these conditions because it not only protects the remaining healthy cells against free-radical damage, but it also seals the microbleedings in the retina that obscure vision.

Additionally, *in vitro* studies indicate the potency of antioxidants to prevent damage to the very sensitive lipids of the retina. Researchers in Japan found that Pycnogenol was the most effective supplement, protecting these structural elements of the eye far better than grape seed extract or vitamins C and E (Chida et al., 1999).

In these studies with retinal tissue, Pycnogenol was 1.5 times more effective than catechin and at least 10 times more potent than grape seed extract, 40 times more potent than vitamin E, 350 times more potent than vitamin C, and 1,000 times more potent than alpha-lipoic acid.

In earlier *in vitro* studies, Pycnogenol was found to be more effective in protecting the retina than several other antioxidants (Ueda et al, 1996).

Cataracts are associated with aging, and are also caused by free radicals. Sunlight is the main source of free radicals that damage the eye lens. Several epidemiological studies have shown that various antioxidant nutrients reduce the incidence of cataracts. Since

Pycnogenol is water soluble, it can bathe the eye with its powerful combination of antioxidants.

A study by Dr. J. R. Trevithick of the University of Western Ontario in Canada has shown that Pycnogenol helps to prevent cataract development in diabetic rats (Trevithick et al., 2000).

Bilberry for Night Vision

There have been about thirty clinical trials studying the possibility that bilberry improves night vision. The formal studies are inconclusive; some agree with this premise, others refute it. WWII fighter pilots swore by it. In any event, the OPCs in bilberry help protect the carotenoids in the eye macula and reduce the free radicals in the eye that can lead to cataracts and retinopathy.

Several studies support the use of bilberry against diabetic retinopathy, and it is prescribed for this use in Europe. Bilberry extract has also been shown to arrest cataract formation in forty-eight or fifty senile cortical cataracts (one of three main types of age-related cataracts). A good review of bilberry's role in vision is given for physicians in *Alternative Medicine Review* (www.thorne.com/altmed rev/.fulltext/6/5/500.html).

The eyes are directly exposed to strong ultraviolet radiation and require the protection from nature's antioxidants—the blues, violets, and browns of bioflavonoids, and the yellows and reds of the carotenoids. However, the OPC story doesn't end here. OPCs protect from many other diseases and disorders as well. Let's look at a few more in the next chapter.

CHAPTER EIGHT

Helping Protect against More Than Age-Related Diseases

D iabetes mellitus is a disorder in which the body cannot convert foods properly into energy. The damage to the cells that leads to either type 1 ("juvenile") diabetes or type 2 ("adult onset") diabetes may involve free-radical reactions, but once the islets of Langerhans cells of the pancreas (type 1) or cellular mechanisms for utilizing insulin have been damaged, antioxidants cannot reverse this. However, diabetes itself increases the production of free radicals, which further damage the body, increasing the risk of heart attack, nerve damage (diabetic neuropathy), cataracts (diabetic cataract), blindness (diabetic retinopathy), and more complications.

Arthritis is not one disease, but several diseases possibly having several causes. The word "arthritis" is derived from a Greek word meaning "joint" and actually means inflammation of the joint. Inflammation increases the production of free radicals.

As stressed often in this book, these diseases and many other degenerative diseases are either caused by or involve free radicals in their pathology. Here's where powerful antioxidant protection is especially needed—arthritics and diabetics need more antioxidant protection than healthy people. OPCs contain the most powerful antioxidant nutrients known at this time. This chapter will discuss how OPCs can improve not only diabetes and athritis, but also asthma, sexual function, memory function, and many other conditions that while not life-threatening can greatly impact the quality of day-to-day life in those experiencing them.

OPCs and Diabetes

It is my firm conviction and those of other health experts that all diabetics and pre-diabetics should consider taking OPC supplements! In 2004, two studies were published on Pycnogenol® and type 2 diabetes demonstrating the antioxidant's effectiveness in lowering blood sugar (glucose) levels. Previous research determined that the supplement may help manage diabetic retinopathy. According to a clinical study conducted under the leadership of Dr. Liu and published in *Diabetes Care*, scientists discovered that type 2 diabetes patients had lower blood sugar and healthier blood vessels after supplementing with Pycnogenol (Liu, 2004).

The open, controlled, dose-finding study demonstrated that patients with mild type 2 diabetes, subscribing to a regular diet and exercise program, were able to significantly lower their blood sugar levels when they supplemented with Pycnogenol. A dosage as low as 50 mg (milligrams) significantly lowered blood sugar and 100 mg further lowered blood sugar levels, whereas higher dosages only marginally further increased the effect.

A second clinical study published in 2004 shows that type 2 diabetes patients who continued to take their antidiabetic medication further lowered blood sugar levels and increased cardiovascular function after supplementing with Pycnogenol. The double-blind, placebo-controlled study published in the October 2004 issue of *Life Sciences*, found that seventy-seven type 2 diabetics who supplemented with 100 mg of Pycnogenol for twelve weeks, during which their standard antidiabetic treatment was continued, significantly lowered blood glucose levels as compared to a placebo (Liu, 2004).

"To the 17 million Americans living day to day with type 2 diabetes, this news is a welcome sign of the safe and effective natural, alternative dietary supplement choices available to them like Pycnogenol," said Dr. Peter Rohdewald, at the Institute of Pharmaceutical Chemistry, University of Muenster, and one of the authors of the study.

Additional related studies of the benefits of Pycnogenol on diabetics were published in *Chinese Pharmacological Bulletin* (Zhang,

2003) and in the *Journal of Biochemical and Molecular Toxicology* (Maritim, 2003; Berryman, 2004).

Pycnogenol also delays sugar absorption. A study published in the *Journal of Diabetes Research and Clinical Practice* reported that Pycnogenol delays the uptake of glucose from a meal more than prescription medications, preventing the typical high-glucose peak in the bloodstream after a meal. The study revealed that pine bark is more potent for suppressing carbohydrate absorption in diabetes than synthetic prescription alpha-glucosidase inhibitors.

"Diabetes mellitus type 2 is a serious disease with rising prevalence," said Dr. Petra Högger, a lead researcher of this study. "This study is crucial for those suffering with the disease because it affirms that Pycnogenol is more effective than [the] prescription medication Precose and supports the abundance of other research done on Pycnogenol and diabetes."

The study was conducted at the University of Würzburg in Germany. Dr. Högger investigated the interaction of Pycnogenol with alpha-glucosidase, an intestinal enzyme that is involved in the digestion of complex carbohydrates such as starch and normal table sugar. Alpha-glucosidase breaks down carbohydrates in a meal into glucose molecules, which are then absorbed into the bloodstream. Results revealed Pycnogenol is 190 times more potent for inhibition of alpha-glucosidase than the synthetic alpha-glucosidase inhibitor Precose, a common prescription medication for treatment of type 2 diabetes (sold in Europe under the name Glucobay).

"The high concentration of procyanidins (flavonoids) found in Pycnogenol is responsible for demonstrating these excellent results," said Dr. Högger. According to Dr. Högger, the large procyanidin molecules were found to be particularly active for inhibiting the activity of alpha-glucosidase, thus demonstrating such notable results. "The carbohydrates enter the bloodstream steadily over prolonged periods of time, which make meals last longer and prolong satiety" (*Diabetes Research and Clinical Practice,* 2007).

Pycnogenol is important for diabetics also because it reduces diabetic microangiopathy. Diabetic microangiopathy is responsible for major diabetic health complications, such as leg ulcers, kidney failure, and retinopathy. It is imperative diabetics receive the best treatment to manage this condition.

For diabetics at risk of leg ulcers, which can lead to amputation, Pycnogenol has been shown to reduce leg ulcers in patients who suffer from diabetic leg ulcerations. A study published in the July 2006 issue of *Clinical and Applied Thrombosis/Hemostasis* showed that patients treated with oral and topical Pycnogenol displayed a 74.4 percent decrease in ulcer size within six weeks (Belcaro, 2006).

The study randomly assigned thirty diabetic patients at the G. D'Annunzio University at Chieti-Pescara, in Italy, who suffer from severe microangiopathy causing leg ulcerations, into four groups. Group 1 participants received 150 mg of Pycnogenol orally and 100 mg topically, as powder applied to the ulcerated area. Group 2 participants received only the oral Pycnogenol, while group 3 received only the topical treatment. Group 4 received no medical care other than routine ulcer care, which was a cleaning with warm water and a local disinfectant.

After six weeks of treatment, the most significant ulcer healing was seen in patients given the combined oral and local treatment, who experienced a 74.4 percent decrease in leg ulcer size. Group 2 patients taking only the oral Pycnogenol experienced a 41.3 percent decrease in leg ulcer size, and group 3 patients using only the topical Pycnogenol experienced a 33 percent decrease in leg ulcer size. Group 4, the control group, experienced only a 22 percent decrease from daily disinfection of the ulcers. Overall, 89 percent of the patients treated with both oral and topical Pycnogenol were completely healed.

According to Dr. Gianni Belcaro, the lead researcher of the study, the majority of diabetic leg amputations (common to the lower leg and feet) begin with the formation of skin ulcers. Impaired blood circulation in diabetics may cause tissue death and discoloration, which leads to the development of ulcers; the open ulcer is then prone to infection and has difficulty healing.

"If left untreated, damage to blood vessels from diabetes then manifests in typical circulatory problems such as hypertension, from which 50 percent of type 2 diabetics suffer," he said. "Solid evidence shows that Pycnogenol effectively reduces high blood pressure, platelet aggregation, LDL cholesterol, and enhances circulation."

Another study led by Dr. Belcaro and published in the September 2006 issue of *Angiology* shows a significant reduction of diabetic microangiopathy (DM) in patients after supplementing with Pycnogenol.

"Diabetic microangiopathy is not a rare phenomenon and essentially affects every diabetic person. The condition may result in vision loss in diabetic retinopathy, kidney problems, and ischemic tissue necrosis causing leg ulcers which may lead to amputation," said Dr. Gianni Belcaro. "With DM, the walls of very small blood vessels (capillaries) become so weak, bleeding and protein leaks occur, which ultimately slows down blood flow, resulting in blood clots and swelling of the limbs (edema)."

The study sampled sixty diabetic patients suffering from DM being treated with insulin for at least three years at G. D'Annunzio University. In addition to their insulin treatment, patients received 150 mg of Pycnogenol orally daily for one month. The control group, 50 percent of the sample, received a placebo. Measurements of blood flow were performed by laser Doppler.

Measurements were taken when patients were lying down and standing up. The capillary adaptation to increased pressure from lying down to standing is generally impaired, due to vessel failure and an increase of pressure in capillaries for individuals who suffer from DM.

Results showed that when patients were lying down, Pycnogenol treatment improved capillary blood flow by 34 percent, compared to 4.7 percent in the placebo group. When patients' blood flow was measured in a standing position, Pycnogenol treatment improved capillary blood flow by 68 percent, compared to 8 percent in the placebo group.

Leakage from capillaries was recorded by measuring ankle swelling, which develops ten minutes after passing from lying down to standing up. After Pycnogenol treatment, swelling was 17 percent lower, compared to 2.6 percent in the placebo group. "The rapid improvement of microvessel complication with Pycnogenol in just four weeks is clinically remarkable," said Dr. Belcaro, who has been involved in a large part of previous Pycnogenol and diabetes related studies.

Earlier studies with more than 1,000 diabetes patients showed that Pycnogenol has the ability to seal leaky capillaries in the eye. This capability stops the progression of vision loss in patients suffering from diabetic retinopathy, a diabetes-induced eye disease that ultimately leads to blindness.

Please keep in mind that diabetics are at increased risk for heart problems and that OPCs help protect against heart disease, as discussed earlier in Chapter 2.

OPCs and Arthritis

Inflammation is characterized by swelling, pain, localized heat, and redness. It can occur due to irritation or injury. Fluid gets trapped in the spaces between cells in the injured tissue. This fluid most often is the result of leakage from capillaries, but it can also be produced directly in tissue via free-radical reactions. Inflammatory cells migrate into inflamed tissue and produce excessive amounts of free radicals. OPC-containing Pycnogenol inhibits accumulation of inflammatory cells, and reduces the output of inflammatory substances (Bayeta and Lau, 2001). Pycnogenol helps normalize capillary permeability to prevent the leakage of fluid that causes edema (swelling). It also helps by neutralizing free radicals that promote swelling and inflammation (Blazso et al., 1994, 1995).

Reducing free radicals eases the swelling associated with inflammation and improves the arthritic condition. A common and possibly the most dangerous free radical called a "superoxide" is involved in the inflammation of arthritis. This was demonstrated by the fact that injections of superoxide dismutase, an antioxidant enzyme that quenches the superoxide free radical, reduced the swelling and inflammation. Experiments by several investigators have shown that Pycnogenol also quenches superoxide free radicals (Packer et al., 1999).

The lead researcher of a recent study of Pycnogenol's beneficial effects, Dr. Petra Högger of University of Würzburg in Germany remarked, "Inflammation is a double-edged sword for our health. Beneficial inflammation is crucial for fighting infections and healing wounds. Harmful inflammation, such as that triggered by a non-infectious event, erroneously aims at the body's tissue, causing significant damage. Patients who supplemented with the pine bark extract Pycnogenol benefited from an immune system response that attenuated excessive inflammation."

"A typical example of harmful inflammation is an asthma attack," said Dr. Högger. "Immune cells in the bronchi perceive harmless substances as foreign, provoking an inflammation response. The inflammation does not have infectious materials to attack so it turns on tissue, causing swellings in the bronchi and greatly impairing breathing. Pycnogenol proved effective at preventing this kind of bad inflammation."

Dr. Högger's study showed that a 200 mg, daily, oral intake of Pycnogenol lowered the activity of nuclear factor-kappa B (NF-kB), the immune cell "trigger" for inflammation, in a group of seven healthy volunteers.

Blood samples were taken prior to the start of the study and again after five days of supplementation. Key immune cells in the blood plasma, also known as "monocytes," drawn after Pycnogenol supplementation showed a significantly lowered NF-kB response by 15 percent, as compared to those without Pycnogenol supplementation.

"This study demonstrates Pycnogenol's ability to inhibit NF-kB and the pro-inflammatory molecules under its control. This reduces 'friendly-fire' incidents where the body's immune system turns inflammation on tissue," said Dr. Högger.

NF-kB is activated by any incident in the body that indicates tissue may have become harmed, such as immune cells recognizing foreign materials. The NF-kB switches on the production of pro-inflammatory molecules required for recruiting immune cells from the bloodstream to the affected tissue.

These immune cells then migrate through blood vessel walls into tissue, where they unload major quantities of destructive enzymes and toxic substances. Different enzymes then dissolve connective fibers such as collagen and elastin to allow immune cells to maneuver freely and to facilitate subsequent tissue healing.

In separate studies published in 2004 and 2006, Pycnogenol demonstrated its anti-inflammatory effects in clinical trials for osteoarthritis, asthma, and dysmenorrhea (menstrual cramps).

Results of the 2006 study, led by Dr. Högger, showed Pycnogenol reduced inflammation relating to symptoms of knee osteoarthritis (OA). The study of OA patients who supplemented with Pycnogenol reported significantly reduced pain and stiffness and increased physical function.

Dr. Högger also found out how Pycnogenol reduces the pain associated with of inflammatory conditions such as arthritis. This 2006 study published in *Biomedicine & Pharmacotherapy* demonstrated the natural pain-reduction properties of Pycnogenol by cyclooxygenase (COX) inhibition. Patients who supplemented with Pycnogenol demonstrated a decrease in pain and inflammation by lowering cyclooxygenase activity, also known as COX-1 and COX-2. COX enzymes produce hormonelike substances called prostaglandins that are vital for proper function of numerous physiologic processes in the body. They are also involved in many pathological processes that are responsible for inflammation and pain. The clinical study investigated the inhibitory effects of Pycnogenol involved in lowering pain in inflammatory conditions such as arthritis.

Two groups of patients were supplemented with Pycnogenol after an initial basal blood sample. The first group of patients was supplemented with Pycnogenol for five consecutive days when another blood sample was taken. The patients' blood samples revealed a direct inhibitory effect on cyclooxygenase activity. Another group of patients was supplemented with a single dose of Pycnogenol to determine how quickly the inhibitory effect on COX enzymes was measurable. Only thirty minutes after ingestion of Pycnogenol, blood samples indicated a statistically significant increase in the inhibition of both COX-1 and COX-2. Dr. Petra Högger stated that the findings suggest that Pycnogenol supplementation inhibits the enzymes involved in the development of pain associated with inflammatory disorders such as arthritis, asthma, and dysmenorrhea.

OPCs and Asthma

Asthma is a lung disease that causes obstruction of the air passages. Asthma is the result of chronic inflammation, which activates immune cells recruited to the bronchi, which in turn causes swellings that greatly impair breathing ability. Asthmatics experience periods of wheezy breathing and breathlessness with intervals of relative or complete freedom from symptoms. Antigenic substances such as pollen and animal hair can obstruct airways, as can chemical irritants like tobacco, smoke, dust, and air pollution. Even cold air and exercise can trigger asthmatic episodes.

A few studies have been carried out with Pycnogenol against asthma. Dr. Ron Watson and his colleagues at the University of Arizona at Tucson found Pycnogenol to help asthmatics. In 2004, Dr. Benjamin Lau and his colleagues at Loma Linda University in California carried out a double-blind, placebo-controlled study with children, aged six to eighteen years, giving excellent results. Study participants who supplemented with Pycnogenol showed a significant reduction of inflammatory mediators (leukotrienes) that cause inflammation and bronchi constriction commonly associated with asthma. Pycnogenol improved pulmonary function, and significantly decreased asthma symptoms, and dramatically lowered the need for using rescue inhalers with albuterol. This is exciting news for parents. All other investigated parameters, such as peak expiratory flow, symptom score, and blood leukotriene levels were also significantly better than in the placebo group. The study was published in the November/December issue of the *Journal of Asthma*.

"Pycnogenol's antioxidant activity and powerful anti-inflammatory properties work to soothe the irritation that causes bronchi to constrict and swell, making breathing difficult," said Dr. Lau, one of the authors of the study. "Treatment compliance is an ongoing problem among the four to five million children under the age of eighteen suffering from asthma. This study demonstrates help for asthmatic children through a nutritional approach as compared to sole reliance on oral medication and rescue inhalers."

As mentioned, Dr. Lau's group tested Pycnogenol's effectiveness based on four parameters: ability to breathe, severity of asthma symptoms, frequency of rescue inhaler usage, and quantity of leukotriene molecules in the child's body. Breathing improved after only one month and continued with further treatment. The severity of asthma symptoms also decreased the longer the children supplemented with Pycnogenol. Likewise, Pycnogenol reduced leukotriene values after one month and further decreased them throughout the three-month study. Pycnogenol dramatically reduced, and in several cases eliminated, the children's dependence on a rescue inhaler, which is used to rapidly dilate the bronchi during an asthma attack.

"Previous studies have shown Pycnogenol to be effective in decreasing asthma symptoms among adults. These recent results further demonstrate the efficacy of Pycnogenol and position this natural antioxidant as a key player in the management of mild to moderate childhood asthma," said Dr. Lau (Lau, 2004).

OPCs and Sexual Function, Fertility, PMS, and Menstrual Disorders

By now it should be no surprise to learn that sexual function and dysfunction can be affected by free radicals as well. Perhaps these problems don't seem as important to many readers as the killer diseases, but for those suffering from infertility or other such disorder, they can be more important.

OPCs can improve fertility. Horse breeders swear by Pycnogenol, if that's any indication of its ability to promote fertility. Sperm is extremely susceptible to oxidative stress. Antioxidants in general improve sperm motility and mobility. Especially useful is Pycnogenol, but vitamin C, selenium, alpha-lipoic acid, and vitamin E are also important. Dr. Scott Roseff and colleagues at the West Essex Center for Advanced Reproductive Endocrinology in West Orange, New Jersey, found that 200 mg of Pycnogenol taken daily for ninety days increased the percentage of structurally normal sperm—that is, non-deformed sperm—by an average of 99 percent. Sperm count did not change. They suggest that this 99 percent increase in structurally normal sperm may allow couples diagnosed with certain types of infertility to forgo *in vitro* fertilization in favor of less invasive and less expensive fertility-promoting procedures (Roseff and Gulati, 1999).

Pycnogenol is also valuable for improving sexual function. In Chapter 2 on protecting the heart and improving circulation, I briefly mentioned nitric oxide's vital role in penile erection. In order for an erection to occur, additional blood must flow into the penis. The arteries supplying this blood depend on nitric oxide to allow them to relax and permit additional blood flow. This nitric oxide is made in the lining of the arteries by the enzyme nitric oxide synthase using the amino acid L-arginine. Pycnogenol can stimulate the production of this enzyme and thus increase production of the needed nitric oxide.

The drug Viagra also works through a mechanism that increases nitric oxide production in these arteries. Clinical studies are available to demonstrate that Pycnogenol is effective in erectile dysfunction. Men who are concerned about erectile dysfunction may want to try supplementing their diet with both OPCs and L-arginine for several weeks (*Journal of Sex & Marital Therapy*, 2003; Muchova, J., Proceedings. Abstract No L 61, 4th International Conference Vitamins 2004 Targeted Nutritional Therapy). Pycnogenol has been awarded a patent (U.S. Patent No. 6,565,851) for relieving the symptoms of erectile dysfunction with OPCs.

There have been significant studies with Pycnogenol revealing treatment efficacy of common problems associated with menstruation, such as dysmenorrhea and menstrual pain. Pycnogenol has also been granted a patent (U.S. Patent No. 6,372,266) for reducing PMS, and menstrual pain and discomfort.

Pycnogenol also significantly reduces endometriosis, a condition in which the tissue that normally lines the uterus (endometrium) develops into growths or lesions in other areas of the body, causing pain, irregular bleeding, and possible infertility. Endometriosis affects women in their reproductive years and is estimated to affect over one million women in the United States. It is one of the common reasons women have to undergo hysterectomies and laparoscopic surgery. The average diagnostic age is twenty-five to thirty, however, endometriosis has been reported in girls as young as eleven years of age.

Research has shown a reduction in abdominal pain due to endometriosis. Studies have shown a clear improvement in terms of reduction of menstrual cramps and pain in 73 percent of women following administration of 30 mg Pycnogenol a day for one month, in addition to those with endometriosis. Abdominal pain due to endometriosis was reduced in 80 percent of the patients, and cramps disappeared in 77 percent of the women taking Pycnogenol according to a study published in the *European Bulletin of Drug Research* (Kohama, 1999).

A study published in the March 2007 issue of the *Journal of Reproductive Medicine* reveals that Pycnogenol significantly reduces symptoms of endometriosis by 33 percent. "The cause of endometriosis is unknown and treatment to fully cure endometriosis has yet to be developed," said Dr. Takafumi Kohama, a lead researcher of the study. "Common hormone treatments such as gonadotropin-releasing hormone agents (Gn-RHa) may likely restrict women from becoming pregnant during treatment. Danazol, another hormone treatment, produces side effects such as ovarian deficiency, osteoporosis, and obesity. Our results convey Pycnogenol as an extremely effective natural treatment without dangerous side effects," he said.

The study, held at Kanazawa University School of Medicine in Ishikawa, Japan, sampled fifty-eight women ages twenty-one to thirty-eight, who underwent operations for endometriosis within six months prior to the study. After confirming regular menstruation and ovulation for three months before treatment, patients were examined before, and at four, twelve, twenty-four, and forty-eight weeks after treatment began, to check for symptom control (pain, urinary and bowel symptoms, breakthrough bleeding, etc.). Pain was evaluated by patients' self-assessment and an investigator interviewed and performed a gynecologic examination.

Patients were randomized into two groups: Pycnogenol and Gn-RHa. Patients who supplemented with Pycnogenol took 30 mg capsules orally twice daily for forty-eight weeks immediately after morning and evening meals. Patients who received the Gn-RHa therapy received injected leuprorelin acetate (a Gn-RH analog), 3.75 mg intracutaneously, six times every four weeks for twenty-four weeks. (Leuprorelin treatment completely blocks estrogen in the body and must be discontinued after twenty-four weeks because of potential side effects.)

Both treatment groups showed no differences at the start of treatment and reported severe pain, pelvic tenderness, and pelvic indurations (abnormally hardened tissue). After four weeks, Pycnogenol slowly but steadily reduced all symptoms from severe to moderate. Treatment with Gn-RHa reduced the scores more efficiently, but after twenty-four weeks post-treatment a relapse of symptoms occurred.

"As expected, Gn-RHa suppressed menstruation during treatment, whereas no influence on menstrual cycles was observed in the Pycnogenol group. Gn-RHa lowered estrogen levels drastically and, in contrast, the estrogen levels of the Pycnogenol group showed no systematic changes over the observation period," said Dr. Kohama. "In addition, five women in the trial taking Pycnogenol actually got pregnant," he said.

When Gn-RHa is used continuously for more than two weeks, the production of estrogen stops, depriving the endometrial implants of estrogen, causing them to become inactive and degenerate. Most women will stop bleeding within two months of starting treatment and return of ovulation and menstruation varies.

OPCs and Memory

Pycnogenol has even been shown to improve brain function and memory according to a study published by Dr. Benjamin Lau and his colleagues at Loma Linda University, as discussed earlier in Chapter 4. If you remember (humor intended), Dr. Lau found Pycnogenol offset the age-related declines in memory retention and learning.

Other studies have demonstrated how effectively Pycnogenol shields brain cells called neurons or neuronal cells from oxidative stress. These cells help process and transmit information within the body. Dr. Dave Schubert of the Salk Institute of Biological Studies in San Diego found that Pycnogenol is so powerful that it completely prevents oxidative damage caused by beta-amyloid, a protein associated with Alzheimer's disease. Dr. Lester Packer of the University of California, Berkeley, has shown that Pycnogenol protects neuronal cells from damage associated with excessive amounts of the flavor enhancer glutamate (Kobayashi et al., 2000).

The nervous system consists of very fragile cells and for optimum performance they depend on excellent protection, as provided by Pycnogenol. We need all our mental capacity to stay competitive in the world today, where the ability to retain and recall knowledge and information are the keys to success. Pycnogenol helps us to keep the pace.

Living Better, Longer

So, it's not just an increased lifespan or the protection from the killer diseases that OPCs have to offer—they can help protect our quality of life. OPCs help us keep a sharper mind, avoid dementia, save our eyesight, maintain a healthy sex life, and reduce the chances of our getting a debilitating disease such as diabetes, asthma, and arthritis. Not bad for a group of natural nutrients.

CHAPTER NINE

Relieving Allergies

OPCs are helpful against diseases and conditions other than those that are age-related or even free-radical related. Decades before free radicals were known to damage the body and the exciting news about the many roles of OPCs in reducing the risk of the killer diseases, such as cancer and heart disease, were known, OPCs were successfully used—especially in Europe—to control hay fever and other allergies.

The Allergic Reaction

Allergies are hypersensitive reactions that occur when the body comes in contact with harmless substances that it perceives as harmful. Substances that cause these reactions are termed allergens. When a hypersensitive person comes into contact with an allergen, the body releases histamine in an attempt to fight off the allergen. This release of histamine triggers the symptoms so common to allergies—inflammation, sneezing, runny nose, and itchy eyes.

Effects of OPCs on Allergies

OPCs help control symptoms associated with allergic reactions in several ways. At this time, none is understood as being free-radical related, but the following mechanisms are known.

First, OPCs appear to block histamine release in *in vitro* studies. Professor Sharma from the department of pharmacology of the University of Dublin has shown that OPCs inhibit the secretion of histamine by specialized white blood cells called mast cells. This action is mediated through the free-radical scavenging property of OPCs. These findings were presented during the British Pharmacology Society meetings held in Dublin in July 2001, and results were published in a peer-reviewed journal in 2003 (*Phytotherapy Research*). Antihistamines generally work by interfering with the attachment of histamine to cells in the linings of the throat and nose after it has been released. It's more efficient to prevent histamine's release in the first place than to try to keep released histamine away from its receptors on target cells.

Second, OPCs appear to increase the uptake and re-uptake of histamine into histamine's own storage granules, where it's out of the way and can't cause misery.

What may prove to be a third important mechanism was reported by Dr. David White of the University of Nottingham in England. Dr. White's studies suggest that OPCs inhibit the production of histamine. They block the action of an enzyme called histidine decarboxylase, which forms histamine from the amino acid histidine.

OPCs may be effective against allergies without producing side effects such as drowsiness and dry mucous membranes often caused by conventional allergy medications.

CHAPTER TEN

Treating Attention Deficit Disorder

It's impossible for me to write about the health benefits of OPCs without mentioning the research on attention deficit disorder (ADD), or as it is sometimes called attention deficit hyperactivity disorder (ADHD). This disorder used to be known simply as "hyperactivity."

Attention deficit disorder involves the inability to keep focused on a task, impulsive behavior, and/or hyperactivity. ADD is much broader than the excessive physical activity of hyperactivity; it includes behavioral and mental disorders that keep one from learning or performing well, even though the individual has the mental capability to do so.

ADD affects about 5 to 10 percent of school-aged children in the United States, and is the basis for about one-half of the childhood referrals to diagnostic clinics. ADD is seen ten times more frequently in boys than girls.

Understanding ADD

The cause of ADD is not known, but structural abnormalities have been ruled out by CAT, MRI, and EEG scans (computerized axial tomography, magnetic resonance imaging, and electroencephalogram, respectively). The leading suspect appears to be neurotransmitter abnormalities, possibly associated with decreased activity or stimulation in the upper brainstem and frontal-midbrain. There is also suspicion that toxins, environmental problems, or neurologic immaturity could be involved.

The American Psychiatric Association lists fourteen signs of which at least eight must be present to be officially classified as ADD. These fourteen signs are:

1. Often fidgets with hands or feet, or squirms in seat (restlessness);

2. Has difficulty remaining seated when required to do so;

3. Is easily distracted by extraneous stimuli;

4. Has difficulty awaiting turn in games or group activities;

5. Often blurts out answers before questions are completed;

6. Has difficulty in following instructions;

7. Has difficulty sustaining attention in tasks or play activities;

8. Often shifts from one uncompleted task to another;

9. Has difficulty playing quietly;

10. Often talks excessively;

11. Often interrupts or intrudes on others;

12. Often does not seem to be listening to what is being said;

13. Often loses things necessary for tasks or activities;

14. Often engages in physically dangerous activities without considering possible consequences.

Different combinations of inattentiveness or impulsiveness or hyperactivity constitute different sub-groups of ADD. Note that ADD is not a form of retardation, although learning under these conditions is difficult.

Not Just Kids

Usually, ADD begins by four to seven years of age, with the peak ages in which treatment is sought is between eight and ten years of age. However, adolescents and adults also suffer from ADD, with the estimated incidence ranging from 30 to 70 percent of the affected children also becoming ADD adults. In adults, ADD is rarely a case of "adult onset," but an abating childhood condition. The physical "hyperactivity" lessens with age, but the adult still has marked attention problems and can be impulsive. Problems in the adolescent and adult occur predominantly as academic failure, low self-esteem, and difficulty learning appropriate social behavior. Often they have personality-trait disorders, antisocial behavior, short attention spans, poor social skills, and are impulsive and restless.

Treatment Problems

The conventional treatment for ADD is with central nervous system stimulants such as amphetamines. Ritalin (methylphenidate or methyl alpha-phenyl-2-piperidineacetate) is widely prescribed, as a consequence 5 to 10 percent of our youngsters are going to school drugged. According to the *Merck Manual* (Merck, 2006), common side effects of Ritalin are sleep disturbances (e.g., insomnia), depression or sadness, headache, stomachache, suppressed appetite, elevated blood pressure, decreased learning, behavioral changes, and reduction of growth.

The long-term benefits of Ritalin have not been conclusively demonstrated.

What is known is that there are risks in taking the ADD drugs. Thousands of patients are admitted to hospitals each year due to ADD-drug problems. A Roanoke, Virginia, high-school senior died while "snorting" Ritalin—the current drug of choice among high school students.

Pycnogenol for ADD

In February 1995, I received a letter from a Valdosta, Georgia, elementary school teacher for seventeen years, who is the mother of five children including an ADD daughter. I emphasize that she is a teacher because teachers know what is normal behavior in the classroom. The mother was then giving Pycnogenol to her daughter and had been able to take her off Ritalin completely. She was concerned about safe dosage limits and said she would appreciate any information I could give her regarding ADD and Pycnogenol. At that time, I had nothing unique to tell her about ADD and Pycnogenol. Later that year, I received a call from Myles Lipton of St. Louis. I had met Myles in 1994. Myles owned a health food store and wanted to know if I had any information on ADD and Pycnogenol. He had several adult customers who belonged to a ADD support group, and they were markedly improved.

I often get reports of how Pycnogenol helps this or that, but these "studies of one" are considered as anecdotal reports and have little scientific value. But, when you get enough smoke, you begin looking for the fire. Myles had just presented an opportunity to improve my database regarding ADD and Pycnogenol.

ADD support groups are somewhat of a rarity. ADD sufferers aren't fond of meetings, and it is difficult to conduct such meetings. Myles was familiar with several of the group members and the progress that was being made. He accepted the challenge to get good data. He designed a questionnaire and contacted members of the group.

ADD Study Excerpts

Interviews were conducted at the ADD support group meeting on April 13, 1995. A follow-up interview revealed that only Pycnogenol produced the dramatic results reported here. The interviews were conducted separately so that none influenced any other.

Dan

Dan H. was a thirty-year-old male who always had trouble concentrating and had not performed well in school as a child. He was always very agitated.

When asked why he was taking Pycnogenol, he responded. "I attended ADD meetings for three months, and in speaking with other members, I heard lots of complaints of side effects of various drugs that their doctors had prescribed. I did not want to take these drugs. I was talking with a guy at work and he told me about Pycnogenol."

Dan started taking Pycnogenol. The first day he took 300 mg (milligrams) and then dropped to 150 mg daily. Within days he noticed that he was thinking clearer. His thinking got sharper and continued to improve for a week and then leveled off. He used to be hyper, but was now calmer and more alert to what was going on about him. He reports being less depressed and has less of a sense of being anxious.

Dan reports, "How I have felt in the last two months has been the best thing that has ever happened to me. I am undoing old bad habits, the worst of which was a lot of procrastination in taking care of things that I needed to do. I do a better job of listening to people and when they talk to me, my mind does not drift off anymore."

Jill

Jill G. was a thirty-six-year-old female. She had a short attention span, a short-fused temper, and was always fidgety, impatient, and a master procrastinator.

When asked why she started taking Pycnogenol, she responded, "Dan, a member of our group, came to a meeting and seemed much different. He used to never sit still; he'd interrupt others frequently, and his eyes always seemed like they were popping out of their sockets. But this time, he was sitting still and attentively listening to others, and his eyes looked normal. I asked him what was up and he told me that he was taking 150 mg of Pycnogenol a day."

Jill began taking 150 mg of Pycnogenol daily. She noticed a significant improvement in about two weeks. At first there was an improvement in bowel function. (Zoloft, the antidepressant she was taking, may make one constipated, and she cut back on it.) She noticed a major improvement in the dark circles under her eyes. She seemed to have more motivation and did not procrastinate as much.

Jill adds, "I had been taking Zoloft and have gotten off of it with consultation with my doctor. I have noticed that my husband, Jeff, who also has ADD, has had major improvements and a noticeable increase in confidence since he has been taking Pycnogenol."

Jeff

Jeff G., Jill's husband, was a thirty-five-year-old male who had always been compulsive, had a lot of trouble concentrating, and was always losing things.

When Jeff was asked why he was taking Pycnogenol, he responded, "Because of the radical positive behavioral changes in Dan, a member of our group."

Jeff had begun taking 180 mg of Pycnogenol daily about a month ago, at the same time his wife started. He noticed a marked improvement in about three hours! He went to the bar that he frequents but always feels uncomfortable in. But after three hours of taking Pycnogenol, he felt more at ease. As time went by, the improvements increased, he had less anxiety feelings, more energy, was calmer, and had better relationships with fellow employees at work.

Jeff adds, "I had been taking medication that had the side effect of increasing nervousness, and Pycnogenol has allowed me to reduce the medication and have an overall better feeling."

David

David L. was thirty years old. All through school he was hyperactive and had problems concentrating. David had begun taking 75 mg of Pycnogenol a day about three to four weeks earlier, soon after he noticed the changes in Dan. David noticed positive results after one week. He had less anxiety, felt calmer, and was thinking more clearly.

When asked if he had anything to add, David said, "Just that I really feel better since taking Pycnogenol and I hope that the positive results continue."

Laura

Let's leave the ADD group and return to the Georgia teacher's daughter for an update. Laura was now twelve years old. She has been diagnosed as having ADHD and oppositional and defiant behavior disorder. She had been taking 70 mg of Ritalin and 50 mg of the antidepressant Tofranil daily. Within thirty days of starting both Pycnogenol and a combination antioxidant formula, her parents noted improvements in both Laura's behavior and attention span to a greater extent than that realized with the prescription drugs. Additional improvements in appetite and personality traits were noted.

Beth, Laura's mother, told me, "As a parent, I would never revert to the prescription drug Ritalin. As a mother of five children and an elementary teacher for seventeen years, I have had a wealth of experience in working with children with ADD/ADHD. Most of these children have been on various prescription drugs of which Ritalin, in various dosages, being the most common. It appears that the use of prescription drugs for this disorder has become common place to the point that it is overprescribed. The long-term side effects or ramifications of using drugs like these described are still unknown. The ability to use a food supplement and obtain like, positive results must be pursued."

Since then several important clinical studies have been published showing that Pycnogenol can provide relief of hyperactivity and improve attention in children with ADD/ADHD. To learn more, see *European Child and Aldolescent Psychiatry* (Trebatická, 2006); *Journal of the American Academy of Child & Adolescent Psychiatry* (Heimann, 1999); and *Attention Deficit Disorder* (Hanley, Impact Communications, 1999).

Conclusion

M y purpose in writing this book was to inform you of research and nutrients to help you live better, longer. I have told of the many years that I have spent researching antioxidant nutrients and a few discoveries I have made along the way. This research has been exciting for me, and it represents my life's professional work. The reason I wanted to share my excitement with you is so that you can decide if you wish to take advantage of not only my research, but the research of hundreds of my colleagues as well.

I encourage you to read the original scientific reports by researchers, including Drs. Denham Harman, Lester Packer, Bruce Ames, Ronald Watson, Peter Rohdewald, Frank Schoenlau, Albert Szent-Györgyi, and so many others who conduct research with antioxidant nutrients. An easy way to find their research articles is via the computerized database kept by the National Library of Medicine at www.ncbi.nlm.nih.gov/sites/entrez?db=pubmed and at www.pubmed.gov. Here you can find abstracts to most all articles that I mentioned in this book, and for many, you can find the complete text.

Hopefully, these articles will also encourage you to eat a good diet, and to scientifically fortify this good diet with dietary supplements, especially good antioxidant supplements, including OPCs.

The evidence that I have presented in this book explains how important OPCs are and how they will benefit your health. The option to live better, longer is available. The choice is yours. Be well!

Selected References

Bagchi D., M. Bagchi, S.J. Stohs, et al. "Free radicals and grape seed proanthocyanidin extract: Importance in human health and disease prevention." *Toxicology* 148 (2–3); (Aug 2000): 187–197.

Bayard, V., F. Chamorro, J. Motta, N.K. Hollenberg. "Does flavanol intake influence mortality from nitric oxide-dependent processes? Ischemic heart disease, stroke, diabetes mellitus, and cancer in Panama." *International Journal of Medical Sciences* 4 (Jan 2007): 53–58.

Berryman, A.M., A. Maritim, R.A. Sanders, R.A. et al. "Inhibitory effect of Pycnogenol® on generation of advanced glycation end products *in vitro*." *Journal of Biochemical and Molecular Toxicology* 18 (2004): 345–352.

Bomser, J., D.L. Madhavi, K. Singletary, et al. "In vitro anticancer activity of fruit extracts from *Vaccinium* species." *Planta Medica* 62 (Jun 1996): 212–216.

Csiszar, A., K. Smith, N. Labinskyy, et al. "Resveratrol attenuates TNF-alpha-induced activation of coronary arterial endothelial cells: Role of NF-kappaB inhibition." *American Journal of Physiology—Heart and Circulation Physiology* 291; 4 (2006): H1694–H1699.

Dolara P., C. Luceri, C.D. Filippo, et al. "Red wine polyphenols influence carcinogenesis, intestinal microflora, oxidative damage and gene expression profiles of colonic mucosa in F344 rats." *Mutation Research* 591 (Dec 11, 2005): 237–246.

Fremont, L., M.T. Gozzelino, and A. Linard. "Response of plasma lipids to dietary cholesterol and wine polyphenols in rats fed polyunsaturated fat diets." *Lipids* 35 (Sept. 2000): 991–999.

Giovannelli, L., G. Testa, C. De Filippo, et al. "Effect of complex polyphenols and tannins from red wine on DNA oxidative damage of rat colon mucosa *in vivo*." *European Journal of Nutrition* 39 (Oct 2000): 207–212.

Halpern, M.J., A. Dahlgren, I Laakso, et al. "Red-wine polyphenols and inhibition of platelet aggregation: Possible mechanisms, and potential use in health promotion and disease prevention." *Journal of International Medical Research* 26 (Aug–Sept 1998): 171–180.

Hanley, J.L., "Attention deficit disorder: Pycnogenol® is recommended for treatment of Attention Deficit Disorder. Green Bay, WI: Impact Communications Inc., 17–19.

Heimann, S.W. "Pycnogenol® for ADHD?" *Journal of the American Academy of Child & Adolescent Psychiatry* 38 (1999): 357–358.

Hosseini, S., J. Lee, R.T. Sepulveda. "Pycnogenol® in the management of asthma." *Journal of Medicinal Food* 4 (2001): 201–209.

Kamei, H., Y. Hashimoto, T. Koide, et al. "Anti-tumor effect of methanol extracts from red and white wines." *Cancer Biotherapy and Radiopharmaceuticals* 13 (Dec 1998): 447–452.

Kohama, T. and N. Suzuki. "The treatment of gynaecological disorders with Pycnogenol®." *European Bulletin of Drug Research* 7 (1999): 30–32.

Lau, B.H.S., S.K. Riesen, K.P. Truong, et al. "Pycnogenol® as an adjunct in the management of childhood asthma." *Journal of Asthma* 41(2004): 825–832.

Liu, X., H.J. Zhou, and P. Rohdewald. "French maritime pine bark extract Pycnogenol® dose-dependently lowers glucose in type 2 diabetic patients." 27 (Mar 2004): 839.

Liu, X., J. Wei, F. Tan, et al. "Pycnogenol®, French maritime pine bark, improves endothelial function of hypertensive patients." *Life Sciences,* 74 (2004): 855–867.

Lotito, S.B., B. Frei. "Consumption of flavonoid-rich food and increased plasma antioxidant capacity in human: Cause, consequence, or epiphenomenon?" *Free Radical Biology and Medicine* 41 (Dec 2006): 1727–1746.

Maritim, A.C., B.A. Dene, R.A. Sanders, et al. "Effect of Pycnogenol® treatment on oxidative stress in streptozotocin-induced diabetic rats." *Journal of Biochemical and Molecular Toxicology* 17 (2003): 193–199.

Mink, P.J., C.G. Scrafford, L.M. Barraj, et al. "Flavonoid intake and cardiovascular disease mortality: A prospective study in postmenopausal women." *American Journal of Clinical Nutrition* 85 (March 2007): 895–909.

Muchova, J., Z. Chovanova, M. Hauserova, et al. "The effect of natural polyphenols (extract from *Pinus pinaster* (Pycnogenol®) and ginkgo biloba (EGB 761) on the oxidative stress and erectile function in patients suffering from erectile dysfunction." Proceedings. (Abstract No L 61) 4th International Conference Vitamins 2004 Targeted Nutritional Therapy, Sept 13–15, 2004, Aula Univerzity Pardubice.

Murphy, M. "Vitamin rethink on the cards?" *Chemistry & Industry* (12 March 2007): 5.

Nash, D.T., S.D. Nash, W.D. Grant. "Grapeseed oil: A natural agent which raises serum HDL levels." *Journal of the American College of Cardiology* (1993) 925–116.

Passwater, R.A. "Cancer: New Directions." *American Laboratory* 5; 6 (1973): 10–22, and *International Laboratory* (July/Aug 1973).

Passwater, R.A. and P.A. Welker. "Human Aging Research, Part I." *American Laboratory* 3; 4 (1971): 36 40 and *International Laboratory* (May/June 1971): 24–28.

Passwater, R.A. and P.A. Welker. "Human Aging Research, Part II." *American Laboratory* 3; 5 (1971): 21–26 and *International Laboratory* (July/Aug 1971): 37–40.

Schafer, A. and P. Högger. "Oligomeric procyanidins of French maritime pine bark extract (Pycnogenol®) effectively inhibit alpha-glucosidase" *Diabetes Research and Clinical Practice* 77 (July 2007): 41–46.

Seeram, N., L. Adams, Y. Zhang, et al. "Blackberry, black raspberry, blueberry, cranberry, red raspberry, and strawberry extracts Iinhibit growth and stimulate apoptosis of human cancer cells *in vitro*." *Journal of Agricultural and Food Chemistry* 54 (Dec 13, 2006): 9329–9339.

Shen M., G.L. Jia, Y.M. Wang, et al. "Cardioprotective effect of resveratrol pretreatment on myocardial ischemia-reperfusion induced injury in rats." *Vascul Pharmacol*, 45 (2006):122–126.

Shuguang, L., Z. Xinwen, X. Sihong, et al. "Role of Pycnogenol® in aging by increasing the *Drosophila's* life-span." *European Bulletin of Drug Research,* 11 (2003): 39–45.

Singletary, K.W. and B. Meline. "Effect of grape seed proanthocyanidins on colon aberrant crypts and breast tumors in a rat dual-organ tumor model." *Nutrition and Cancer* 39 (2001): 252–258.

Stanislavov, R. and V. Nikolova. "Treatment of erectile dysfunction with Pycnogenol® and L-arginine." *Journal of Sex & Marital Therapy* 29 (May–June 2003): 207–213.

Trebatická, J., S. Kopasová, Z. Hradecná, et al. "Treatment of ADHD with French maritime pine bark extract, Pycnogenol®." *European Child and Adolescent Psychiatry* 20 (2006): 1–7DOI 10.1007/s00787-006-0538-3 (online version).

Ye, X., R.L. Krohn, W. Liu, et al. "The cytotoxic effects of a novel IH636 grape seed proanthocyanidin extract on cultured human cancer cells." *Molecular and Cellular Biochemistry* 196 (June 1999): 99–108.

Zhang, T.M., C.H. Han, Y.M. Han, et al. "Inhibitory effect of Pycnogenol® on generation of advanced glycation end products in vitro." *Chinese Pharmacological Bulletin* 19 (2003): 437–440.

Index

About the Author

Richard A. Passwater, Ph.D., is a biochemist, whose laboratory research led to the discovery of antioxidant synergism in 1962, which has been the focus of his research and numerous patents ever since. He has authored more than 45 books and booklets as well as more than 400 articles on nutrition and nutritional supplements. Some of his well-known books include *Supernutrition: Megavitamin Revolution* (Dial Press, 1975), *Cancer Prevention and Nutritional Therapies* (McGraw-Hill, 1998), and *The User's Guide to Pcynogenol* (Basic Health, 2005). His books have been translated into ten languages; Spanish, French, German, Chinese, Japanese, Italian, Portuguese, Hebrew, Russian, and Swedish.

Dr. Passwater served on the editorial board of the *Journal of Applied Nutrition*. He is currently the scientific editor for *Whole Foods* magazine, for which he writes a monthly column.

Dr. Passwater's discoveries have led to his recognition worldwide. He has received many awards and recognitions for his work. He was twice honored by the Committee for World Health (1978, 1980), was elected a fellow of the American Institute of Chemistry (1984), received the President's Award for the National Nutritional Foods Association (1999), and was recipient of the John Peter Zenger Free Press Award (2004) and the prestigious James Lind Scientific Achievement Award for his contribution to the scientific advancement of natural medicine (2004). He is listed in the *Who's Who in America*, *Who's Who in the World*, and *Who's Who in the Frontiers of Science*.

For additional details about Dr. Passwater and his work, visit www.drpasswater.com.